SpringerBriefs in Applied S
and Econometrics

SpringerBriefs present concise summaries of cutting-edge research and practical applications across a wide range of fields. Featuring compact volumes of 50 to 125 pages, the series covers a range of content, from professional to academic. Briefs are characterized by fast, global electronic dissemination, standard publishing contracts, standardized manuscript preparation and formatting guidelines, and expedited production schedules.

SpringerBriefs in Applied Statistics and Econometrics (SBASE) showcase areas of current relevance in the fields of statistical methodology, theoretical and empirical econometrics, statistics in the natural sciences, and financial econometrics. Of interest are topics arising in the analysis of cross-sectional, panel, time-series and high-frequency data. The primary audience of the SBASE series are researchers with an interest in theoretical, empirical and computer-based statistics and econometrics.

The SBASE series accepts a variety of formats:

- timely reports of state-of-the art techniques,
- bridges between new research results, as published in journal articles, and literature reviews,
- snapshots of a hot or emerging topic,
- lecture or seminar notes making a specialist topic accessible for non-specialist readers.

Manuscripts presenting original results or novel treatments of existing results are particularly encouraged. All volumes published in SBASE undergo a thorough refereeing process.

The SBASE series is published under the auspices of the German Statistical Society (Deutsche Statistische Gesellschaft).

More information about this subseries at http://www.springer.com/series/16080

Aygul Zagidullina

High-Dimensional Covariance Matrix Estimation

An Introduction to Random Matrix Theory

Aygul Zagidullina
Department of Economics
University of Konstanz
Konstanz, Germany

ISSN 2524-4116 ISSN 2524-4124 (electronic)
SpringerBriefs in Applied Statistics and Econometrics
ISBN 978-3-030-80064-2 ISBN 978-3-030-80065-9 (eBook)
https://doi.org/10.1007/978-3-030-80065-9

Mathematics Subject Classification: 62-02, 62H20, 62H12, 62H10, 62H15, 62H25, 15B52, 60B20

1st edition originally published electronically on KOPS of the University of Konstanz, 2019

This Springer imprint is published by the registered company Springer Nature Switzerland AG.
The registered company address is: Gewerbestrasse 11, 6330 Cham, Switzerland

To Nicole

Foreword

Statistics, as much as any other science, is characterized by a continual and ongoing accumulation of knowledge. As it is difficult to keep up with the subject in its entire breadth, an ever-increasing specialization in the field is unavoidable. Witness the growing body of statistics field journals as well as the rising number of research articles they publish. Because of this development, there is a need for publication outlets that address a broad readership and present challenging material, both new and established, in a self-contained and accessible manner, without requiring specialist background knowledge.

This is the backcloth against which the SpringerBriefs in Applied Statistics and Econometrics (SBASE) were conceived by the German Statistical Society (the *Deutsche Statistische Gesellschaft, DStatG*) and Springer. The general aim of the SpringerBriefs format, i.e., reports of state-of-the-art techniques, snapshots of emerging topics, or lecture notes making a specialist topic accessible to non-specialist readers, seemed particularly well-suited to address the aforementioned need for manuscripts that are too long for being journal articles and too short as traditional monographs.

We were very grateful for being appointed as the inaugural editors of SBASE. This mandate is both a responsibility and an opportunity to establish this new series as a trusted source of high-quality research. We assembled to that end a superb team of associated editors who, with their expertise and network, provide invaluable support in the refereeing process of manuscripts: H. Peter Boswijk (Universiteit van Amsterdam), Jörg Breitung (Universität zu Köln), Walter Krämer (Technische Universität Dortmund), Karl Mosler (Universität zu Köln), and Peter X. K. Song (University of Michigan). With their help, we are confident that we will edit many exciting projects on statistical and econometric methodology or empirical applications in the natural, social, and economic sciences. We hope that the installments of SBASE will find many interested readers among researchers, practitioners, and policymakers alike.

The present book by Aygul Zagidullina is a worthy first volume of SBASE. Its topic, viz., covariance matrix estimation and random matrix theory, straddles traditional statistics as well as a state-of-the-art technique for dealing with high-dimensional data. The book is thus a timely introduction to a topic that will be of interest to readers in economics, finance, engineering, and the medical sciences. The text will be of particular use to those who have a background in classical statistical inference but seek guidance for dealing with data structures that require a different approach to distribution theory. Aygul Zagidullina provides a lucid and insightful bridge between these challenging strands of the literature.

Lastly, we would like to thank Veronika Rosteck and Gerlinde Schuster, our wonderful contacts at Springer, for their competent and patient help. And we are indebted to Wolfgang Schmid, former president and now vice president of the DStatG, for his incessant support and invaluable advice. Without him, SBASE would not have seen the light of the day.

Vallendar, Germany
Dresden, Germany
May 2021

Michael Massmann
Ostap Okhrin

Acknowledgments

I would like to thank Michael Massmann, Ostap Okhrin, the editor, the associate editor, and the anonymous referee for their valuable comments and suggestions that helped to improve the exposition of the book.

I also want to thank the entire team of *SpringerBriefs in Applied Statistics and Econometrics* for its superb work leading to the book's final publication.

Furthermore, I am grateful to my Ph.D. supervisor, Winfried Pohlmeier, for his support and guidance throughout writing this book. I am also grateful to Lyudmila Grigoryeva for her advice and insightful feedback.

I want to thank further Lucas Burger and Julia Heigle for their assistance with tables and figures.

Finally, I would like to express gratitude to my parents and husband for their love and encouragement over the years.

Partial financial support from the German Academic Exchange Service (DAAD) is gratefully acknowledged.

General Conventions and Notation

Let X_n be an $n \times p$ matrix whose entries X_{ij} are i.i.d. with $E(X_{ij}) = 0$ and $\mathrm{Var}(X_{ij}) = 1$, where n is the number of observations and p is the number of variables.

The data matrix $Y_n = X_n \Sigma_n^{1/2}$ is observed, while X_n and Σ_n are not observed, where Σ_n is a $p \times p$ population covariance matrix. The subscript n in Σ_n emphasizes the fact that the population covariance matrix Σ_n depends on n through the dimensionality $p = p(n)$.

Let $\{(\tau_{n,1}, \ldots, \tau_{n,p});\ (v_{n,1}, \ldots, v_{n,p})\}$ denote a system of eigenvalues and eigenvectors of the population covariance matrix Σ_n.

Let $\{(\lambda_{n,1}, \ldots, \lambda_{n,p});\ (u_{n,1}, \ldots, u_{n,p})\}$ denote a system of eigenvalues and eigenvectors of the sample covariance matrix $\widehat{S}_n$ of Y_n, where the sample covariance matrix is defined as:

$$\widehat{S}_n = \frac{1}{n} Y_n' Y_n = \frac{1}{n} \Sigma_n^{1/2} X_n' X_n \Sigma_n^{1/2}.$$

In what follows, the set of eigenvalues $\tau_{n,1}, \ldots, \tau_{n,p}$ or $\lambda_{n,1}, \ldots, \lambda_{n,p}$ is referred to as the *spectrum* of Σ_n or $\widehat{S}_n$, respectively.

An empirical distribution function of the population eigenvalues is defined as following:

$$H_n(x) = \frac{1}{p} \sum_{i=1}^{p} \mathbb{1}\{\tau_{n,i} \le x\},\ x \in \mathbb{R}^+.$$

An empirical distribution function of the sample eigenvalues is, respectively:

$$F_n(x) = \frac{1}{p} \sum_{i=1}^{p} \mathbb{1}\{\lambda_{n,i} \le x\},\ x \in \mathbb{R}^+.$$

The empirical distribution functions, $H_n(x)$ and $F_n(x)$, defined in terms of the spectrum of Σ_n or $\widehat{S}_n$, respectively, are called empirical *spectral distributions*.

Thus, $H(x)$ is the limiting population spectral distribution corresponding to the sequence $\{H_n(x)\}$, and $F(x)$ is the limiting sample spectral distribution corresponding to $\{F_n(x)\}$.

$\|A\|$, $\|A\|_F$, and $\|A\|_1$ denote the operator, Frobenius, and the l_1-norm of a matrix A, respectively. They are defined as $\|A\| = \sqrt{\lambda_{\max}(A'A)}$, $\|A\|_F = \sqrt{\mathrm{tr}\,(A'A)}$, and $\|A\|_1 = \max_j \sum_i |a_{ij}|$.

Contents

Chapter 1
Introduction

This book is concerned with covariance matrix estimation. In particular, we focus on the sample covariance matrix estimator and introduce related aspects of the random matrix theory. Although this matrix is commonly used in many fields of study and its properties have long been known in the context of classical statistics, recent advances in computer science and data access have revived interest in studying sample covariance matrix from the random matrix theory perspective. This alternative framework provides powerful tools that enable the analysis of random matrices (and sample covariance matrix, in particular) stemming from the data that researchers and practitioners currently encounter.

At present, fields such as economics, finance, genomics, atmospheric sciences, wireless communications, and biomedical imaging must account for an increasing number of variables in the available data. This data can be characterized as high-dimensional, which means the number of observed covariates (features) is comparable to, or exceeds, the available observations. Such a setting poses a challenge to the traditional multivariate statistical theory that evolved under different assumptions to analyze datasets that consist of a small number of variables.

Historically, multivariate statistical tools were developed based upon an assumption that the observations are independent and have multivariate normal distributions. These conditions allow the closed-form expressions to be derived for the finite sample distributions of certain statistics that are often used in the inference. By weakening the above requirements, one can also derive the distributions of interest; however, this can only be achieved asymptotically (with an increasing number of observations) and for the particular moment restrictions. Thus, the classical statistical inference was mainly devised in a setup where the dimension p is fixed and the number of samples n grows to infinity. This is the so-called fixed p, large n setting.

This fixed dimension assumption, which the traditional statistical inference builds upon, is difficult to justify in the context of Big Data, which implies a significant number of variables.

A. Zagidullina, *High-Dimensional Covariance Matrix Estimation*,
SpringerBriefs in Applied Statistics and Econometrics,
https://doi.org/10.1007/978-3-030-80065-9_1

In particular, the limitations of "fixed p, large n" setting are apparent when considering the covariance matrix estimation problem, where the number of parameters is of the order $\mathcal{O}(p^2)$, implying that the dimension can easily become larger than the sample size n. Therefore, the traditional asymptotic approximations derived for the covariance matrix estimator may no longer be valid in finite samples and may yield misleading results, even for reasonably large sample sizes.

Furthermore, many multivariate statistical tools, such as likelihood ratio tests (LRT), or dimension-reduction techniques, such as factor analysis (FA) and principal component analysis (PCA) (Jolliffe 2002), are based upon the sample covariance matrix. As a consequence, their finite sample properties and performance can also be negatively affected. Spearman's psychometrics studies on general intelligence are the earliest example of an application closely related to social sciences using the covariance matrix as an input (Spearman 1904). This work is well-known for introducing the statistical notion of FA based on the covariance matrix. Hence, the problem of covariance matrix estimation has always played a key role not only for multivariate analysis but also when inferring the underlying latent structures from the observed data.

Moreover, the sample covariance matrix and its spectral decomposition (i.e., eigendecomposition) have a direct effect on many other multivariate statistical techniques, including the non-linear versions of PCA (such as kernel PCA methods in machine learning applications; e.g., Chapter 14 in Hastie et al. 2009; El Karoui 2010), canonical correlation analysis (CCA) (e.g., Hardoon et al. 2004), multivariate analysis of variance (MANOVA), factor-augmented vector autoregressive (FAVAR) models (Bernanke et al. 2005), large-scale hypothesis testing (i.e., testing the covariance structure, e.g., Anderson 1963; Bai et al. 2009; Jiang 2016), inferring a graphical structure in the network analysis (where the inverse, or precision matrix, is used; e.g., Kolaczyk 2009; Nadakuditi and Newman 2012), and the ordinary least squares (OLS).

Thus, researchers and practitioners encounter the problem of covariance matrix estimation in various contexts and real-world applications. Although the sample covariance matrix estimator has a conceptually simple definition and is easy to compute, it is essentially the matrix-valued random variable, or the random matrix, and as such, it has a more complex nature. Therefore, the finite sample and limiting properties of this estimator should be studied within the theory developed for the random matrices.

The classical statistical random matrix theory (RMT) was pioneered by J. Wishart. In Wishart (1928), he derived the finite sample distribution for the unnormalized sample covariance matrix, $n\widehat{S}_n$, which is, however, only valid under the normality assumption imposed on the matrix entries and in the case when the dimension is smaller than the sample size. This finite sample distribution for the random matrix $n\widehat{S}_n$ is known as a Wishart distribution.

In contrast, for the non-Gaussian or distribution-free assumption imposed on the matrix entries, the distributional result is derived only at the limit. Under the mild moment conditions imposed on the data, the standardized covariance matrix estimator (demeaned and rescaled) has an asymptotic multivariate normal distribution.

The above classical RMT results were established for the sample covariance matrix; however, in general, RMT is not only concerned with the behavior of matrices, but also with the properties of corresponding eigenvalue statistics and eigenvectors. This naturally follows from the fact that (random) matrices are intrinsically linked to their eigenvalues, or the spectrum. Thus, the spectrum behavior mostly governs the properties of random matrices, both in finite samples and under the asymptotic regime. Therefore, studying the properties of spectral statistics is of utmost importance when analyzing random matrices (e.g., Bai & Silverstein 2010).

In the traditional "fixed p, large n" setting, the sample eigenvalue estimators are consistent and unbiased, so the sample covariance matrix also possesses these desirable properties, although only at the limit. In small samples, however, the performance of traditional eigenvalue estimators degrades.

It has long been known that the largest sample eigenvalues are upward-biased, while the smallest ones are downward-biased (see, Friedman 1989), and moreover, this bias becomes more pronounced with a decreasing number of observations. As a result, the sample covariance matrix also performs poorly. For normally distributed data, the theoretical explanation for this well-known fact was established in Lawley (1956). Lawley derived the asymptotic expansions for the mean and variance of the traditional eigenvalue estimators, which as a byproduct, demonstrated an explicit form of finite sample bias.

Thus, the traditional asymptotic results do not approximate the finite sample behavior of the eigenvalue estimators (respective eigenvectors) with sufficient accuracy. As a consequence, they are not informative with regard to the properties of the sample covariance matrix when the dimension is comparable to sample size. More importantly, the traditional asymptotics do not allow the dimension to grow in proportion to the sample size and, therefore, is not appropriate in finite samples when approximating the high-dimensional setting in which both p and n are large or p is larger than n.

The characteristics of sample eigenvalues, or a sample spectrum, and of the respective sample covariance matrix observed in the finite samples are, however, well explained and approximated for finite p and n by the high-dimensional asymptotics, when both p, $n \to \infty$, such that their dimension-to-sample ratio $c_n = p/n \to c > 0$. This is the so-called large p, large n setup, and is why the research in a field of random matrices turned to this form of asymptotics as an alternative to the traditional.

The modern RMT currently used to study the properties of the spectrum, and hence the random matrices under high-dimensional asymptotics, was initiated by E. Wigner in the late 1960s as a branch of nuclear physics theory. However, at present, it is important in many other research fields, including modern econometric analysis. This is an advanced and technical branch of the statistical literature but is also inspiring, as many state-of-the-art asymptotic results derived in the context of RMT approximate the behavior of estimators (and test statistics) precisely for a dimension and sample size as small as five (see, e.g., Johnstone 2001).

Furthermore, the bound and distributional results for the spectral statistics derived under high-dimensional asymptotics can potentially be exploited by researchers and practitioners in many other fields of study and areas of application. Moreover, the high-dimensional approximations are of great importance for modern data analysis in the era of Big Data. Therefore, the aims of this survey are as follows: (1) To introduce in a broadly understandable and non-technical way the basic concepts and major results related to spectral statistics and random matrices (with a particular focus on the sample covariance matrix) under the high-dimensional asymptotics; (2) to provide an overview of the analogous results that are valid under traditional asymptotics; and (3) to connect two statistical frameworks, traditional and high-dimensional. This book is intended for a broad readership that may include applied econometricians and machine learning practitioners and thus has two objectives. The first is to draw attention to the deficiencies of standard statistical tools when used in the high-dimensional setting. The second is to provide insight into recent RMT developments and hence to inspire future researchers to integrate these developments into their work when analyzing high-dimensional data.

The survey primarily studies the properties of the sample covariance matrix, and its respective eigenvalues and eigenvectors, under two asymptotic regimes. The limiting characteristics of these estimators, however, are heavily dependent on the population quantities. Therefore, throughout the book, we concentrate on the theoretical results for the two types of possible population designs for the covariance matrix: the full-rank and the reduced-rank structures.

The full-rank structure of the population covariance matrix is a prime example of traditional and high-dimensional asymptotics, although in the high-dimensional setup, when there are no "isolated," extreme eigenvalues, this structure is called "*the null case*."

The alternative reduced-rank covariance structure is also of great importance under both asymptotic regimes. This structure is naturally linked to the factor models and the dimension-reduction techniques, as it is assumed that the original data, in that case, is generated by a few latent factors. This assumption is appealing in many fields of study, particularly in economics and finance (see, e.g., Chamberlain & Rothschild 1983). In the traditional "fixed p, large n" setting, this design is referred to as the covariance matrix with the underlying factor structure.

In contrast, in the "large p, large n" setting, when a few population eigenvalues are much larger than the majority (or in other words, are spiked), the reduced-rank structure of the covariance matrix is called "*the non-null case*" or "*spiked covariance model*" (the latter term was introduced in Johnstone 2001).

This factor structure design of the covariance matrix, or the presence of a few spiked population eigenvalues, is a compelling model of reality in many applications. Furthermore, it implies the intrinsically low-dimensional representation of the original data and is crucial to detect the reduced-rank structure of the population covariance matrix in the context of dimension-reduction techniques. Therefore, in this survey, we also concentrate on the sphericity and partial sphericity tests commonly used in the traditional multivariate statistical theory to discover the reduced-rank—or the factor structure—in the population covariance matrix.

To explain the above topics best and to fulfill our objectives, we have decided to organize the book as follows. Chapter 2 briefly reviews the basic terms, concepts, and major results for traditional estimators (the sample covariance matrix, its eigenvalues, and eigenvectors) in the standard "fixed p, large n" framework and introduces the sphericity and partial sphericity tests. Hence, we begin by considering the classical RMT results for the sample covariance matrix and its spectral decomposition, where "classical" is used, as earlier, to indicate that the results for the random matrix -covariance matrix estimator—are derived under the traditional asymptotics. The discussion in this chapter is crucial for the understanding of how the results derived in the classical framework differ from or are similar to the corresponding findings provided in the high-dimensional RMT framework (the latter is introduced and discussed extensively in Chap. 4). Thus, Chap. 2 is essential to clearly represent these seemingly different branches of literature as two parts of a whole.

After discussing the theoretical results, Chap. 3 illustrates the finite sample properties of these estimators and test statistics when p is of the same order as n and demonstrates the deficiencies of the traditional limiting results when applied to finite samples.

Next, Chap. 4 highlights RMT developments for the sample estimators and mainly concentrates on the limiting results, while the finite sample results are omitted for concision. The non-asymptotic results for the random matrices and the matrix concentration inequalities are discussed extensively in Vershynin (2018) (see also Vershynin 2012). Furthermore, as mentioned above, the RMT-based limiting results closely approximate the finite sample behavior of the estimators. Thus, we focus in this chapter on the properties of the Wishart matrices (i.e., the unnormalized sample covariance matrices, $\widehat{S}_n$) and their respective eigenvalues and eigenvectors under high-dimensional asymptotics. However, our discussion here is biased toward the results established for the sample eigenvalues. This bias follows from the fact that many fundamental results in modern RMT have been derived for the spectrum of random matrices. Thus, we concentrate here on the bound and fluctuation results of the spectrum statistics.

As stated earlier, in the high-dimensional setup, we differentiate between two cases when discussing the properties of estimators: the null and non-null scenarios. Moreover, we consider the behavior of a majority of the eigenvalues, referred to in the literature as "*bulk*," separately from the behavior of "*extremes*" (the largest and smallest eigenvalues). To summarize, in Chap. 4, we explain the behavior of the bulk and extremes in both the null and non-null cases. These limiting results are mostly based on the distribution-free assumptions that are imposed on the data.

In the null case, we concentrate on the foundational result of Marčenko and Pastur (1967), which describes the behavior of bulk and has been used as a basis for many of the statistical tools developed in the last decade under high-dimensional asymptotics. Furthermore, we discuss the limiting fluctuation results for the extreme sample eigenvalues: the Tracy–Widom laws for the largest and smallest eigenvalues.

In the non-null case, we consider the recent theoretical findings for the sample eigenvalues generated from the spiked population model Paul (2007) and Bai and Yao (2008) and demonstrate that in this particular case, the high-dimensional asymptotics results can be understood as generalizations of well-known traditional ones. Finally, we discuss the recent corrections of the sphericity and partial sphericity tests under high-dimensional asymptotics, and introduce the alternative testing procedures based on developments in RMT and which have the desired size and power characteristics under high-dimensional asymptotics.

We summarize the discussion and provide an outlook of the literature in Chap. 5. Furthermore, we present all theoretical results that are considered throughout the book in a table (see table in Sect. 5), which provides a holistic view of the findings corresponding to two asymptotic frameworks, thus serving as a further guide throughout the survey.

Although this book introduces recent results in the theory of random matrices, due to the special focus on the properties of sample covariance matrix and its eigendecomposition, as well as space limitations, we do not cover many other related topics. For example, we do not discuss a mathematical theory of the spectrum of random matrices, and in particular, we do not address *Wigner matrices*. The interested reader is referred to the books of Anderson et al. (2009) and Mehta (1990) for a rigorous introduction to the mathematical tools used for the analysis of random matrices (see also a collection of overviews dedicated to a broad range of topics in the book of Akemann et al. 2011). In addition, we do not explain central limit theorems for linear spectral statistics of large-dimensional random matrices and the respective link to the multivariate statistical tests; a strong introduction and thorough treatment of these topics are given in Bai and Silverstein (2010) and Yao et al. (2015). Furthermore, a large branch of literature dedicated to shrinkage estimators of covariance and precision matrices using RMT is not considered in this survey, although we shed light on the motivation of this research direction by presenting the theoretical rationale for the sample covariance matrix deficiencies in high dimensions. The results of this literature are vital as they demonstrate how the traditional covariance and precision matrix estimators can be corrected in a high-dimensional setting using the RMT findings. Ledoit and Wolf (2015), Wang et al. (2015), and Bodnar et al. (2016), among others, introduce recent developments in this area. Moreover, we do not consider covariance and precision matrix estimators with sparsity or factor structures (see excellent overview by Fan et al. 2016). Finally, we do not discuss the properties of correlation matrices (and the correlation structure tests), which are closely related to the properties of the covariance matrix; the interested reader is referred to the papers of El Karoui (2009), Gao et al. (2017), Bun et al. (2017), and Zheng et al. (2019). These references are not intended to be comprehensive but rather guide the reader through the enormous body of literature.

References

Akemann, G., Baik, J., & Di Francesco, P. (2011). *The Oxford handbook of random matrix theory*. Oxford University Press.

Anderson, G. W., Guionnet, A., & Zeitouni, O. (2009). *An introduction to random matrices*. Cambridge studies in advanced mathematics. Cambridge University Press.

Anderson, T. W. (1963). Asymptotic theory for principal component analysis. *Annals of Mathematical Statistics, 34*(1), 122–148.

Bai, Z., Jiang, D., Yao, J.-F., & Zheng, S. (2009). Corrections to LRT on large-dimensional covariance matrix by RMT. *Annals of Statistics, 37*(6B), 3822–3840.

Bai, Z., & Silverstein, J. W. (2010). *Spectral analysis of large dimensional random matrices* (2nd ed.). New York: Springer.

Bai, Z., & Yao, J. (2008). Central limit theorems for eigenvalues in a spiked population model. *Annales de l'Institut Henri Poincaré, Probabilités et Statistiques, 44*(3), 447–474.

Bernanke, B. S., Boivin, J., & Eliasz, P. (2005). Measuring the effects of monetary policy: A factor-augmented vector autoregressive (FAVAR) approach. *Quarterly Journal of Economics, 120*(1), 387–422.

Bernanke, B. S., Boivin, J., & Eliasz, P. (2016). Direct shrinkage estimation of large dimensional precision matrix. *Journal of Multivariate Analysis, 146*, 223–236.

Bun, J., Bouchaud, J.-P., & Potters, M. (2017). Cleaning large correlation matrices: Tools from Random Matrix Theory. *Physics Reports, 666*, 1–109.

Chamberlain, G., & Rothschild, M. (1983). Arbitrage, factor structure, and mean-variance analysis on large asset markets. *Econometrica, 51*(5), 1281–304.

Chamberlain, G., & Rothschild, M. (2009). Concentration of measure and spectra of random matrices: Applications to correlation matrices, elliptical distributions and beyond. *The Annals of Applied Probability, 19*(6), 2362–2405.

El Karoui, N. (2010). The spectrum of kernel random matrices. *Annals of Statistics, 38*(1), 1–50.

Fan, J., Liao, Y., & Liu, H. (2016). An overview of the estimation of large covariance and precision matrices. *The Econometrics Journal, 19*(1), C1–C32.

Friedman, J. H. (1989). Regularized discriminant analysis. *Journal of the American Statistical Association, 84*(405), 165–175.

Gao, J., Han, X., Pan, G., & Yang, Y. (2017). High dimensional correlation matrices: The central limit theorem and its applications. *Journal of the Royal Statistical Society. Series B: Statistical Methodology, 79*(3), 677–693.

Hardoon, D. R., Szedmak, S., & Shawe-Taylor, J. (2004). Canonical Correlation Analysis: An overview with application to Learning Methods. *Neural Computation, 16*(12), 2639–2664.

Hastie, T., Tibshirani, R., & Friedman, J. (2009). *The Elements of Statistical Learning, 2nd Edition*, Springer Series in statistics. New York: Springer.

Jiang, D. (2016). Tests for large-dimensional covariance structure based on Rao's score test. *Journal of Multivariate Analysis, 152*, 28–39.

Johnstone, I. M. (2001). On the distribution of the largest eigenvalue in principal components analysis. *Annals of Statistics, 29*(2), 295–327.

Jolliffe, I. T. (2002). *Principal component analysis* (2nd ed.). New York: Springer.

Kolaczyk, E. (2009). *Statistical analysis of network data—methods and models*. Springer series in statistics. New York: Springer.

Lawley, D. N. (1956). Tests of significance for the latent roots of covariance and correlation matrices. *Biometrika, 43*(1–2), 128–136.

Lawley, D. N. (2015). Spectrum estimation: A unified framework for covariance matrix estimation and PCA in large dimensions. *Journal of Multivariate Analysis, 139*, 360–384.

Marčenko, V. A., & Pastur, L. A. (1967). Distribution of eigenvalues for some sets of random matrices. *Mathematics of the USSR-Sbornik, 1*(4), 457–483.

Mehta, M. L. (1990). *Random matrices* (2nd ed.). Academic Press.

Nadakuditi, R. R., & Newman, M. E. J. (2012). Graph spectra and the detectability of community structure in networks. *Physical Review Letters, 108*, 188701.

Paul, D. (2007). Asymptotics of sample eigenstructure for a large dimensional spiked covariance model. *Statistica Sinica, 17*(4), 1617.

Spearman, C. (1904). "General Intelligence," objectively determined and measured. *The American Journal of Psychology, 15*(2), 201–292.

Vershynin, R. (2012). Introduction to the non-asymptotic analysis of random matrices. In *Compressed sensing: Theory and applications* (pp. 210–268). Cambridge University Press.

Vershynin, R. (2018). *High-dimensional probability: An introduction with applications in data science*. Cambridge series in statistical and probabilistic mathematics. Cambridge University Press.

Wang, C., Pan, G., Tong, T., & Zhu, L. (2015). Shrinkage estimation of large dimensional precision matrix using random matrix theory. *Statistica Sinica, 25*(3), 993–1008.

Wishart, J. (1928). The generalised product moment distribution in samples from a normal multivariate population. *Biometrika, 20A*(1–2), 32–52.

Yao, J., Zheng, S., & Bai, Z. (2015). *Large sample covariance matrices and high-dimensional data analysis*. Cambridge University Press.

Zheng, S., Cheng, G., Guo, J., & Zhu, H. (2019). Test for high-dimensional correlation matrices. *The Annals of Statistics, 47*(5), 2887–2921.

Chapter 2
Traditional Estimators and Standard Asymptotics

Abstract We discuss the concept of a random matrix and relate it to the sample covariance matrix estimator. This short review is intended to guide the reader through the classical results.

Keywords Covariance matrix estimator · Wishart matrix · Eigenvalue decomposition · Factor model · Sphericity tests

In this chapter, we briefly recall some basic terms and review major results for the sample estimators of the covariance matrix and its respective eigenvalues and eigenvectors under the traditional asymptotics. We define the concept of a random matrix, relate it to the sample covariance estimator, and discuss the distribution of this matrix-valued random variable in finite samples and under an asymptotic regime. Using a motivating example, we go on to demonstrate that the inferential properties of random matrices are tied to their respective eigenvalues, and thus, the study of random matrices is naturally linked to and builds on the theory developed for eigenvalue estimators. Given this connection, we elaborate on the theory behind the eigenvalue and respective eigenvector estimation. Further, we discuss a particular type of random matrix that naturally arises in many fields of study, such as economics, finance, signal processing, and genomics. These matrices are generated by intrinsically low-dimensional data that has a so-called factor structure. Such structure in the data directly affects the eigenvalues of a random matrix and, hence, its rank, in this case reducing that rank. To detect the reduced rank of a matrix and an implied low-dimensional representation of the data, two tests well-known in multivariate statistics are performed: the sphericity and partial sphericity tests. For this reason, we also focus on the distributional results for the statistics of these tests. In this chapter we mainly present the traditional, well-known results derived in the "fixed p, large n" setup. However, when it is worthwhile, we also refer to the analogous results that are valid in the high-dimensional "large p, large n" setting. We notice that in the traditional setting, the population covariance matrix Σ does not depend on n, as the dimension p is fixed and does not increase with n. Thus, the notation introduced earlier applies mostly to the high-dimensional

A. Zagidullina, *High-Dimensional Covariance Matrix Estimation*,
SpringerBriefs in Applied Statistics and Econometrics,
https://doi.org/10.1007/978-3-030-80065-9_2

framework. To sum up, the results in this section are essential to understanding the changes that statistical properties of sample estimators are subject to under high-dimensional asymptotics. The latter we discuss further in Chap. 4.

2.1 Sample Covariance Matrix

Before discussing sample covariance matrix properties, we briefly define a random matrix. It is simply a matrix that has as its entries independent scalar-valued random variables, and thus, it represents a generalization of the scalar random-variable concept to the higher-dimensional space.

In fact, the sample covariance matrix estimator, $\widehat{S}_n$, is the most important, well-known, and widely used random matrix in statistics and other related fields. This matrix estimator is based on the observed sample Y_n ($n \times p$ matrix of data) and defined as follows:

$$\widehat{S}_n = \frac{1}{n} Y_n' Y_n = \frac{1}{n} \Sigma^{1/2} X_n' X_n \Sigma^{1/2},$$

where $n \times p$ matrix X_n and the true $p \times p$ covariance matrix, Σ, are not known.

From the above definition one can see that the sample covariance matrix is essentially the sum of n random matrices:

$$\widehat{S}_n = \frac{1}{n} \sum_{i=1}^{n} Y_i' Y_i,$$

where each term $Y_i' Y_i$ is a $p \times p$ matrix-valued random variable.

In the traditional setup, where the dimensionality p is fixed and smaller than the number of observations n ($p \leq n$) the sample covariance estimator has an exact distribution that is valid in finite samples. This distributional result for the sample covariance matrix $\widehat{S}_n$ is given in the following proposition (Srivastava & Khatri 1979, Section 3.2):

Proposition 2.1.1 *Let Y_n be an $n \times p$ ($p \leq n$) matrix whose rows are i.i.d. $\mathcal{N}_p(0, \Sigma)$ with Σ positive definite $p \times p$ matrix. Then, $n\widehat{S}_n = Y_n' Y_n \sim \mathcal{W}_p(\Sigma, n)$, where $\mathcal{W}_p(\Sigma, n)$ is a Wishart distribution with mean $n\Sigma$ and n degrees of freedom.*

Here we should emphasize that the above statement holds only under the normality assumption and for $p \leq n$.

The Wishart distribution, or the family of Wishart distributions $\mathcal{W}_p(\Sigma, n)$, defined for symmetric, non-negative definite matrix-valued random variables, was first derived in Wishart (1928) and is the generalization of a chi-square distribution. This distribution and the non-central versions of it are commonly used in the inference to derive the distributions of likelihood ratio test (LRT) statistics that are often represented as the ratio of sample covariance matrix determinants (see, e.g., Kollo & von Rosen 1995).

In the high-dimensional setup when $p \gg n$, this distributional result no longer holds. However, the unnormalized version of the sample covariance matrix $n\widehat{S}_n = Y_n'Y_n$ is often referred to as a Wishart matrix in the literature. The name alludes to the above-stated conventional result, although neither the normality assumption nor the condition $p \leq n$ is required in that setup. One of the fundamental results in the random matrix theory (RMT) that we elaborate on in Chap. 4 is derived for the Wishart matrices under high-dimensional asymptotics when both p and n tend to infinity. Hence, the term Wishart matrix is often used in the corresponding high-dimensional covariance literature, although it does not necessarily mean that the matrix of interest $n\widehat{S}_n$ is Wishart-distributed.

Although the finite sample distribution of the covariance matrix estimator is known and has closed-form representation, the asymptotic result for the standardized random variable $B_n = \sqrt{n}(\widehat{S}_n - \Sigma)$ is also of great importance in multivariate statistics as it forms the basis for large sample inference. The following result is according to Anderson (1963) (Theorem 3.4.4).

Proposition 2.1.2 *Let Y_n be an $n \times p$ matrix with rows are i.i.d. $\mathcal{N}_p(0, \Sigma)$ and $n\widehat{S}_n = Y_n'Y_n$. Then, the limiting distribution of $B_n = \sqrt{n}(\widehat{S}_n - \Sigma)$ is normal with mean 0 and covariances*

$$E(b_{ij}b_{kl}) = \sigma_{ik}\sigma_{jl} + \sigma_{il}\sigma_{jk},$$

where $b_{ij} = [B_n]_{ij}$, and σ_{ik} is the ik-th element of the population covariance matrix Σ.

This proposition constitutes the multivariate central limit theorem and shows that the sample covariance matrix $\widehat{S}_n$ is asymptotically normally distributed with mean Σ as sample size increases. The same result can be extended to the case when the rows of the Y_n matrix are i.i.d. random variables with finite fourth-order moments; however, the covariance matrix of B_n will depend on the fourth-order moments of Y_n.

Thus, for an arbitrary population covariance matrix Σ with non-unit eigenvalues, we get an asymptotic normal distribution of the sample counterpart as the sample size increases. However, for the particular case of an identity population covariance matrix I $(p \times p)$, it is proved that the limiting distribution of $\sqrt{n}(\widehat{S}_n - I)$ is not Gaussian (see, e.g., Anderson 1963, Corollary 13.3.2).

Proposition 2.1.3 *Let $n\widehat{S}_n \sim \mathcal{W}_p(I, n)$. Then, the limiting distribution is*

$$\sqrt{n}(\widehat{S}_n - I) \xrightarrow{d} G(W),$$

where $G(W)$ is the distribution function of the Wigner matrix W $(p \times p)$. The probability density function of $G(W)$ is given by

$$g(W) = 2^{-p/2}\pi^{-p(p+1)/4}\exp\left(-\frac{1}{2}\mathrm{tr}W^2\right),$$

where the Wigner matrix W is defined as follows:

$$W_{ij} \sim \mathcal{N}(0, 1), \quad \sigma_{ij} = \frac{1}{2} \text{ and } \sigma_{ii} = 1.$$

Hence, the sample covariance matrix recentered around the population covariance matrix I $(p \times p)$ and rescaled by $\sqrt{n}$ has the asymptotic distribution that coincides with the distribution of the Wigner matrix.

It should be stressed here that the Wigner matrices appear not only under traditional asymptotics in the context of limiting distribution of a sample covariance matrix. They also play a key role in recent developments related to the eigenvalues of square symmetric matrices under high-dimensional asymptotics (RMT) and, thus, often appear in the corresponding literature together with the Wishart matrices (for a detailed and more technical discussion of the topic we refer to an excellent review by Paul and Aue 2014).

The random matrices and sample covariance estimators, in particular, are intrinsically linked to the corresponding eigenvalues (and eigenvectors), and thus, the theoretical properties of these estimators are naturally interdependent. For example, in the subsequent discussion of the results for sample eigenvalues, we often encounter an assumption of a Wishart-distributed sample covariance matrix in the derivations. This link can be explained both intuitively and formally.

From an intuitive point of view, unlike a vector, which is often a subject of interest in the inference, a matrix is the more complex mathematical concept as it has a much richer structure allowing for specific dependencies between columns or rows. That is why, while the complexity measure of a (random) vector is sparsity or the number of non-zero elements, for a (random) matrix an analogous complexity measure is its rank. By definition, a rank is a number of linearly independent columns of a matrix, and it is given as the number of non-zero eigenvalues. Hence, if the matrix has linearly dependent columns, and as a consequence low rank, i.e., $r \ll p$, where $r = \mathrm{rank}(X_n)$, this is equivalent to the sparsity in the corresponding eigenvalue structure, $\tau = (\tau_1, \ldots, \tau_r, 0, \ldots, 0)_{p\times 1}$. Furthermore, the same sparsity degree or a particular structure of the column vectors implies their linear dependence and thus translates into the sparsity in the eigenvalue basis, or into a reduced rank of a matrix (see, e.g., Rigolett 2015).

This explains why the intricate structures of a (random) matrix and, thus, its properties can be described so well predominantly by the corresponding eigenvalues. This link is especially pronounced in the high-dimensional "large p, large n" setup, where the properties of the Wishart matrices, for example, are solely tied to the spectrum or, in other words, to the set of eigenvalues.

To get a more formal explanation, one can turn to the following illustrative example that is based on the well-known matrix decomposition method.

Example 2.1.1 An arbitrary matrix X of dimension $n \times p$, where n is the number of observations and p is the dimensionality, can be written according to the singular value decomposition (SVD) as

$$X = ATV', \text{ where :}$$

(i) A and V are $(n \times r)$ and $(p \times r)$ matrices, respectively, each of which has orthonormal columns so that $A'A = I_r$ and $V'V = I_r$.
(ii) T is an $(r \times r)$ diagonal matrix with elements $(\tau_1, \ldots, \tau_r)$.
(iii) r is the rank of X.

Suppose the rank of the matrix X is equal to 1, i.e., $r = 1$. Then, matrix X is represented in terms of SVD as $X = \tau_1 a v'$. From this expression, it is obvious that all the columns of this matrix are proportional to each other and, thus, are highly correlated. In this particular case, then, the general structure of the matrix, its rank, the linear dependence of the columns, and the resultant implied high correlation among the variables are predetermined by the one non-zero eigenvalue, τ_1.

All the above directly translates to the similar properties of the empirical covariance matrix, which is given by the following expression: $\frac{1}{n}X'X = \frac{1}{n}\tau_1^2 vv'$. Moreover, one can observe that the low-dimensional representation of the original data X is transferred into the respective low-dimensional covariance matrix representation through the sparsity in the eigenvalue basis.

Following up the above example and our discussion of the importance of eigenvalue estimators in the study of random matrices, we continue below with an overview of sample eigenvalues' properties.

2.2 Sample Eigenvalues

In this section, we present the results for the eigenvalues of a sample covariance matrix. We assume the sample covariance matrix to be Wishart-distributed, i.e., $n\widehat{S}_n \sim \mathcal{W}_p(\Sigma, n)$; thus, the findings are valid only under the normality condition imposed on the sample data. The structure of this discussion is similar to that of the sample covariance matrix discussion—namely, we begin with the finite sample results and then turn to the asymptotic ones.

2.2.1 Exact Distribution

The exact joint distribution of sample eigenvalues $\lambda_{n,1}, \ldots, \lambda_{n,p}$ is of a complex form. Its mathematical representation does not explain the behavior of the density function in finite samples and provides little intuition regarding the interaction of sample and population eigenvalues.

Moreover, the exact marginal distributions of sample eigenvalues (for example, the largest and smallest ones that are used in the Roy's test; see, e.g., Roy 1953) are cumbersome to derive. Consequently, the exact moments of the sample eigenvalues are also hard to obtain.

The following result for the exact joint distribution of the sample eigenvalues is according to Muirhead (1982) (Theorem 9.4.1).

Proposition 2.2.1 *Let $n\widehat{S}_n \sim \mathcal{W}_p(\Sigma, n)$, $n > p-1$. Then, the joint density function of the sample eigenvalues $\lambda_{n,1}, \ldots, \lambda_{n,p}$ can be expressed in the following form:*

$$g(\lambda_{n,1}, \ldots, \lambda_{n,p}) = \left(\frac{n}{2}\right)^{pn/2} \pi^{m^2/2} \frac{(\det \Sigma)^{-n/2}}{\Gamma_p(p/2)\,\Gamma_p(n/2)} \prod_{i=1}^{p} \lambda_{n,i}^{(n-p-1)/2} \prod_{i<j}^{p} (\lambda_{n,i} - \lambda_{n,j})$$
$$\cdot\, {}_0F_0^{(p)}\left(-\frac{n}{2}\Lambda_n, \Sigma^{-1}\right) \quad (\lambda_{n,1} > \lambda_{n,2} > \cdots > \lambda_{n,p} > 0),$$

where $\Lambda_n = \operatorname{diag}(\lambda_{n,1}, \ldots, \lambda_{n,p})$ is the diagonal matrix of sample eigenvalues and

$${}_0F_0^{(p)}\left(-\frac{n}{2}\Lambda_n, \Sigma^{-1}\right) = \sum_{k=0}^{\infty} \sum_{\kappa} \frac{C_\kappa\left(-\frac{n}{2}\Lambda_n\right) C_\kappa\left(\Sigma^{-1}\right)}{k!\, C_\kappa(I)}$$

is a two-matrix ${}_0F_0^{(p)}(\cdot, \cdot)$ hypergeometric function that has an expansion in terms of zonal polynomials $C_\kappa(\cdot)$ (symmetric polynomials in eigenvalues), where κ is defined as a partition of positive integers.

The *hypergeometric function of matrix argument* in the above statement involves a series of *zonal polynomials* and is traditionally used in the "fixed p, large n" framework to express the exact non-central distributions of random matrices and to derive the asymptotic distributions of LRT statistics under the alternative hypotheses. These functions facilitate the mathematical representation of integrals, which are defined in terms of matrix argument and cannot be evaluated in closed form.

In the above proposition, the hypergeometric functions arise when we describe the exact density function of sample eigenvalues. This follows naturally from the fact that the density function of a positive definite matrix can be transformed into the density function of its eigenvalues. For further details on the hypergeometric functions and zonal polynomials, we refer to Chapter 7 in Muirhead (1982).

As we can see from the definition of ${}_0F_0^{(p)}$, the exact distribution of the sample eigenvalues depends on the power series, which converge extremely slowly in the case of large n or for large population eigenvalues Muirhead (1978). Thus, any inference based on sample eigenvalues is infeasible when dealing with the exact distributions.

2.2.2 *Finite Sample Bias*

As stated above, tractable expressions for the exact moments of the eigenvalues of sample covariance matrix $\widehat{S}_n$ are not known, but Lawley (1956) has derived asymptotic expansions for the first and second moments.

Proposition 2.2.2 *Let* $n\widehat{S}_n \sim \mathcal{W}_p(\Sigma, n)$. *If* $\tau_1 > \tau_2 > \cdots > \tau_p > 0$ *and* $\lambda_{n,1} \geq \lambda_{n,2} \ldots \geq \lambda_{n,p}$; *then for a fixed* p, *as* $n \to \infty$, *the mean and the variance can be expanded as follows:*

$$E(\lambda_{n,i}) = \tau_i + \frac{\tau_i}{n} \sum_{l=1, l \neq i}^{p} \frac{\tau_l}{\tau_i - \tau_l} + \mathcal{O}\left(\frac{1}{n^2}\right),$$

$$V(\lambda_{n,i}) = \frac{2\tau_i^2}{n} \left(1 - \frac{1}{n} \sum_{l=1, l \neq i}^{p} \left(\frac{\tau_l}{\tau_i - \tau_l}\right)^2\right) + \mathcal{O}\left(\frac{1}{n^3}\right).$$

Hence, in finite samples the bias term for the mean of the sample eigenvalue $E(\lambda_{n,i})$ is of order $\frac{1}{n}$, while for the variance $V(\lambda_{n,i})$ it is of order $\frac{1}{n^2}$. However, as the sample size grows, the bias term in the mean and the variance itself tends to zero, meaning that asymptotically the sample eigenvalue is a consistent estimator of the population counterpart.

An illustration of the finite sample bias for the sample eigenvalues is given in Fig. 2.1. In our example, we focus on the largest sample eigenvalue $\lambda_{n,1}$ of the arbitrary covariance matrix $\widehat{S}_n$ that has dimension $p = 30$.

The graphs show that with an increasing number of observations n, the mean of the sample eigenvalue $\lambda_{n,1}$ approaches the true value τ_1, while the deviation from the estimated mean gets smaller. Hence, the distribution of the sample eigenvalue $\lambda_{n,1}$ is more localized about the true value for large n. In fact, Fig. 2.1 suggests that in order to achieve the asymptotic result for the sample eigenvalue $\lambda_{n,1}$, we need $n = 100{,}000$ observations, which is not a realistic number in the applied research. Furthermore, from the graph we can infer that the largest sample eigenvalue $\lambda_{n,1}$ is an upward-biased estimator for the true eigenvalue τ_1 in finite samples. This undesirable property has been pointed out several times in the corresponding literature (see, e.g., Friedman 1989). The upward bias of the largest sample eigenvalues implies that the variance of the original data explained through the low-dimensional representation is overestimated, and thus the respective dimension-reduction technique can yield a misleading result, for example, a lower-

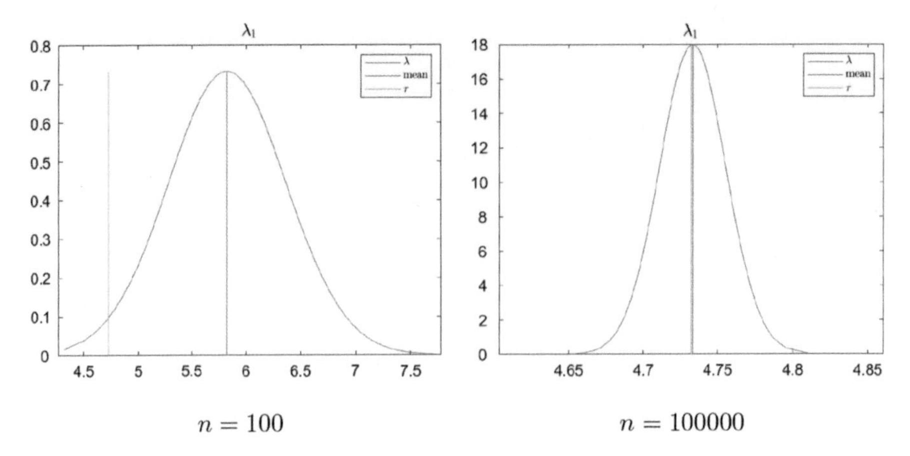

Fig. 2.1 Sample eigenvalues. Finite sample bias. Monte Carlo simulation. The distribution of random variables comprising the $n \times p$ data matrix Y_n is real Gaussian with mean zero and $\Sigma = \text{diag}(\tau_1, \ldots, \tau_p)$. The i-th population eigenvalue is equal to $\tau_i = H^{-1}((i - 0.5)/p)$, $i = 1, \ldots, p$, where the limiting spectral distribution H is given by the distribution of $1 + 10Z$, and $Z \sim Beta(1, 10)$. (This distribution is right-skewed and resembles in shape of an exponential distribution. The support of the distribution is defined on the interval [1,10]. The design is based on Ledoit and Wolf (2015) and approximates very well the pattern of the eigenvalues observed in financial datasets.) $p = 30$

dimensional space is chosen for a projection as compared to the one implied by the true data-generating process. Furthermore, returning to our discussion of the link between eigenvalues and matrices, eigenvalues estimated with a finite sample bias do not reflect the true structure and complexity of the population covariance matrix nor, hence, of the underlying data. This should serve as a cautionary reminder for those engaged in applied research based on sample eigenvalue estimators in finite samples.

In addition to the above, the finite sample bias derived by Lawley (1956) provides intuition about how sample eigenvalues would behave in the high-dimensional setup, where the dimension p is no longer fixed and is assumed to grow together with the sample size n such that their ratio $c_n = p/n$ tends to the finite non-zero constant c as p and n both tend to infinity.

Under the special conditions imposed on the limiting distribution of the sample eigenvalues, and employing the tools from RMT, Mestre (2008) shows that sample eigenvalues are inconsistent under the high-dimensional asymptotics, $p \to \infty, n = n(p) \to \infty, c_n = p/n \to c > 0$. Furthermore, the limiting values of the sample eigenvalues resemble the result of Lawley (1956), which is derived under standard asymptotics with fixed p and $n \to \infty$, albeit with a correction factor c/p:

$$\lambda_{n,i} \to \tau_i + \tau_i \cdot \frac{c}{p} \sum_{l=1, l \neq i}^{p} \frac{\tau_l}{\tau_i - \tau_l} \text{a.s.}$$

Thus, the finite sample bias, which does not affect consistency in the traditional setting, translates into inconsistency of sample eigenvalue estimators under high-dimensional asymptotics.

2.2.3 Asymptotic Distribution

As we have seen earlier, the exact distributional results for sample eigenvalues, as well as their asymptotic approximations (for further details see a review by Muirhead 1978), are of limited use due to the challenging representations.

The more convenient way in this context is to derive the asymptotic distribution of standardized variables. The following result goes back to Anderson (1963) (Theorem 13.5.1).

Proposition 2.2.3 *Let* $n\widehat{S}_n \sim \mathcal{W}_p(\Sigma, n)$. *If* $\tau_1 > \tau_2 > \cdots > \tau_p > 0$ *and* $\lambda_{n,1} \geq \lambda_{n,2} \cdots \geq \lambda_{n,p}$, *then the limiting distribution is*

$$\sqrt{n}(\lambda_{n,i} - \tau_i) \xrightarrow{d} \mathcal{N}(0, 2\tau_i^2),\ i = 1, \ldots, p.$$

Thus, asymptotically the sample eigenvalues are normally distributed. Moreover, the joint distribution of sample eigenvalues is multivariate normal, with the covariances given as

$$\mathrm{Cov}(\lambda_{n,i}, \lambda_{n,i'}) = \begin{cases} 2\tau_i^2, & \text{if } i = i', \\ 0, & \text{if } i \neq i'. \end{cases}$$

Consequently, unlike in the finite samples, asymptotically the sample eigenvalues are independent for normally distributed data. In general, however, this asymptotic independence result does not hold in the high-dimensional setup. This is further discussed in Chap. 4.

To consider the distributional result for sample eigenvalues in the special case of the identity population covariance matrix I $(p \times p)$, we first introduce the eigenvalues of the Wigner matrix, which was defined in Proposition 2.1.3.

Let us denote as $l_1^w \geq l_2^w \geq \ldots \geq l_p^w$ the ordered eigenvalues of the Wigner matrix W.

Proposition 2.2.4 *Let* $n\widehat{S}_n \sim \mathcal{W}_p(I, n)$ *and* $\lambda_{n,1} \geq \lambda_{n,2} \cdots \geq \lambda_{n,p}$. *Then, the limiting distribution is*

$$\sqrt{n}(\Lambda_n - I) \xrightarrow{d} G(l_1^w, l_2^w, \ldots, l_p^w),$$

where Λ_n is the diagonal matrix with $\lambda_{n,1} \geq \lambda_{n,2} \ldots \geq \lambda_{n,p}$ and $G(l_1^w, l_2^w, \ldots, l_p^w)$ is the distribution function of the eigenvalues corresponding to the $p \times p$ Wigner matrix W. The probability distribution function of the eigenvalues is given by

$$g(l_1^w, l_2^w, \ldots, l_p^w) = C_p \exp\left(-\frac{1}{2}\sum_{i=1}^{p}(l_i^w)^2\right)\prod_{i<j}\left(l_i^w - l_j^w\right),$$

where constant C_p is defined as

$$C_p = 2^{-p/2}\pi^{p(p-1)/4}\Gamma_p^{-1}\left(\frac{1}{2p}\right).$$

Hence, the sample eigenvalues centered around the unit population values and scaled by $\sqrt{n}$ behave asymptotically as the eigenvalues of the Wigner matrix (Akemann et al. 2011 (Theorem 28.2.4) and Anderson 1963 (Corollary 13.3.2)).

2.3 Sample Eigenvectors

The eigenvectors of the covariance matrix corresponding to the non-unit population eigenvalues are those vectors pointing in the directions of greatest data variation, or in other words, where the distribution is most extended. The corresponding amount of variation in data is measured by the eigenvalues. This is the common geometric interpretation given to eigenvectors. This is illustrated in Fig. 2.2.

Eigenvectors are essential in PCA and methods that build upon it, as the principal components (PCs) are constructed based on the eigenvectors. The eigenvectors define a linear transformation of the original data, such that the generated PCs are orthogonal to each other given that the first PC has the greatest variance or explanatory power with regard to the original data, the second PC has the second greatest variance, and so on.

Thus, in order to get consistent PC estimates, we need to estimate the eigenvectors consistently. The next proposition shows that under traditional asymptotics the sample eigenvectors are consistent estimators of the population counterparts and, hence, are the respective sample PCs.

Proposition 2.3.1 *Let $n\widehat{S}_n \sim \mathcal{W}_p(\Sigma, n)$. If $\tau_1 > \tau_2 > \cdots > \tau_p > 0$ and $\lambda_{n,1} \geq \lambda_{n,2} \ldots \geq \lambda_{n,p}$, $u_{n,1i} \geq 0$, $v_{1i} \geq 0$, then the limiting distribution is*

$$\sqrt{n}\left(u_{n,i} - v_i\right) \xrightarrow{d} \mathcal{N}(0, \Sigma(\tau_i)),$$

$$\Sigma(\tau_i) = \tau_i \sum_{1\leq k\neq i\leq p} \frac{\tau_k}{(\tau_k - \tau_i)^2} v_k v_k'.$$

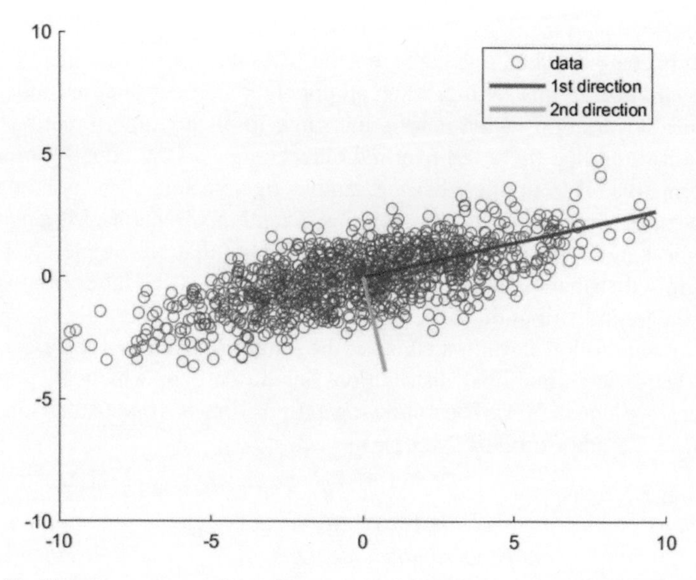

Fig. 2.2 The blue dots represent data points generated from the two-dimensional normal distribution with mean zero and covariance matrix Σ. The red and yellow lines represent the eigenvectors pointing in the directions of greatest variation in the data and second greatest variation, respectively

The conditions $u_{n,1i} \geq 0$ and $v_{1i} \geq 0$ are imposed to rule out sign indeterminacy and to uniquely define the matrices of sample and population eigenvectors, respectively.

According to the above result (Anderson 1963, Theorem 13.5.1), the sample eigenvectors are asymptotically normally distributed. Furthermore, the joint distribution of the sample eigenvectors is multivariate normal with the covariance structure given as follows:

$$\mathrm{Cov}(u_{n,i}, u_{n,i'}) = \begin{cases} \tau_i \sum\limits_{1 \leq k \neq i \leq p} \dfrac{\tau_k}{(\tau_k - \tau_i)^2} v_k v_k', & \text{if } i = i', \\ \\ -\dfrac{\tau_i \tau_{i'}}{(\tau_i - \tau_{i'})^2} v_i v_{i'}', & \text{if } i \neq i'. \end{cases}$$

For a given asymptotic covariance structure of sample eigenvectors, we can see that the eigenvector estimators are not asymptotically independent. Furthermore, the variance of each individual estimator $u_{n,i}$ depends on all remaining population values. And the variation of sample eigenvectors decreases if the population eigenvalues $\tau_1 > \tau_2 > \cdots > \tau_p > 0$ are farther apart from each other. The latter conclusion is implied by the fact that only distinct population eigenvalues have uniquely defined eigenvectors. And given a large number of observations n, these unique eigenvectors can be estimated precisely.

On the contrary, in case of an identity population covariance matrix, the sample eigenvectors are not the consistent estimators any longer. Similarly to the case of eigenvalues (see, e.g., Proposition 2.2.4), the distributional result for the eigenvectors changes when one considers the particular case of an identity matrix for the covariance. In this setting, when all population eigenvalues are unit-valued, the amount of variation in the data is the same in all possible directions. Thus, no direction should be preferred over the others, since all are equally informative or uninformative. This implies that the sample eigenvectors corresponding to the population ones can be considered as pointing in random directions. Moreover, since all have unit length, this is equivalent to the statement that all sample eigenvectors are uniformly distributed on the unit sphere. The proof of the latter proposition is based on the joint distribution of the eigenvectors.

Thus, in what follows, we first consider the joint distribution of the eigenvectors and, afterward, their marginal distributions in the case in which all population eigenvalues are equal to 1. The following proposition is from Anderson (1963) (Theorem 13.3.5 and Corollary 13.3.2):

Proposition 2.3.2 *Suppose $n\widehat{S}_n \sim \mathcal{W}_p(I, n)$, and $\lambda_{n,1} \geq \lambda_{n,2} \dots \geq \lambda_{n,p}$. Then, the matrix of sample eigenvectors $U_n = (u_{n,1}, \dots, u_{n,p})$ has a conditional Haar invariant distribution and is distributed independently of the sample eigenvalues.*

The conditional Haar invariant distribution for the matrix of sample eigenvectors stems from the fact that the entries of this matrix, $u_{n,ij}$s, are independent, and the joint distribution of these entries is invariant under rotation by orthogonal matrices.

The following definition of the conditionally Haar distributed orthogonal matrix is according to Lemma 13.3.2 in Anderson (1963).

Proposition 2.3.3 *If the orthogonal matrix U_n has a distribution such that $u_{n,1i} \geq 0$, and if $U_n^{**} = J(U_n Q')U_n Q'$ has the same distribution for every orthogonal Q, then U_n has a conditional Haar invariant distribution. The matrix J is a diagonal sign matrix such that $J_{ii} = 1$ if $u^*_{n,1i} \geq 0$, and $J_{ii} = -1$ if $u^*_{n,1i} < 0$, where $U_n^* = U_n Q'$.*

Now given that the joint distribution of the eigenvectors is conditional Haar invariant, we can infer the distribution of each eigenvector based on the following useful proposition.

Proposition 2.3.4 *If $p \times p$ orthogonal matrix U_n is Haar distributed, then for any unit vectors $x \in \mathbb{R}^p$, $y = U_n x$ is uniformly distributed over the unit sphere in $\mathbb{R}^p$.*

The above statement means that every sample eigenvector $u_{n,i}$ (for $i = 1, \dots, p$) corresponding to the unit population eigenvalue is uniformly distributed on the sphere defined as $\mathbb{S}^{p-1} = \{x \in \mathbb{R}^p : \|x\| = 1\}$.

The graphical illustration is provided in Fig. 2.3 for the three-dimensional case $p = 3$, with the number of independent draws $R = 1000$.

The above discussion in the PCA context implies that PCs constructed based on sample eigenvectors when the true covariance matrix is identity are randomly directed and, thus, are not informative and have little need for data compression and dimension reduction.

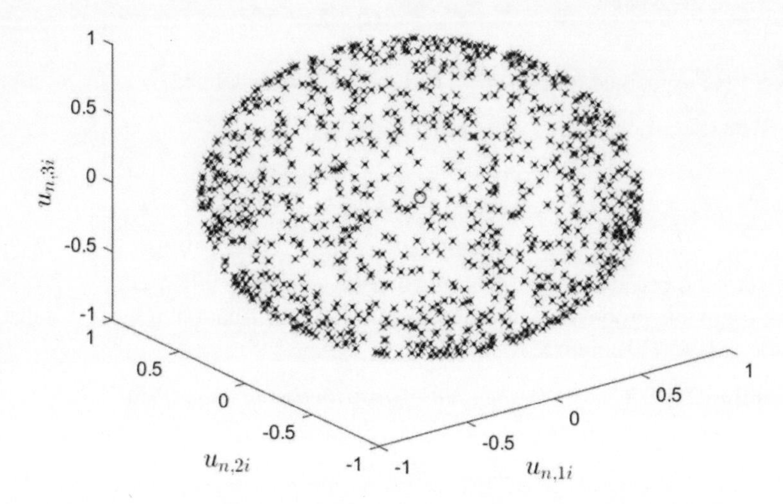

Fig. 2.3 Unit sphere. Distribution of sample eigenvectors. $\Sigma = I$. Monte Carlo simulation

That is why, it is important to detect the identity population structure of the covariance matrix from sample counterpart in order to avoid the misleading inference based on the lower-dimensional data representation. To reveal such a non-informative or noisy feature of the true data process the so-called sphericity test based on the sample covariance matrix is employed in the multivariate statistical analysis. A natural extension of this test to the case when the original data has more pronounced variation in some of the dimensions compared to the others is often referred to as the partial sphericity test.

Thus, continuing the above discussion on the importance of eigenvalues in determining the intrinsically low-dimensional data and its connection to an implied factor structure in the covariance matrix and PCA (or other dimension-reduction techniques), we proceed further with sphericity and partial sphericity tests.

2.4 Sphericity and Partial Sphericity Tests

The sphericity test is used in order to test the null hypothesis that the population covariance matrix is equal to an identity matrix or is proportional to it, i.e., $H_0 : \Sigma = \tau^* I$, and thus, to determine whether there is a possibility for a dimension reduction of the observed data. Furthermore, its test statistic is constructed based on the sample eigenvalues of the covariance matrix.

Thus, the null and alternative hypotheses for the sphericity test are given as follows:

$$H_0 : \Sigma = \tau^* I \text{ vs. } H_1 : \Sigma \neq \tau^* I,$$

where τ^* is unspecified (unknown). The LRT statistic is defined by

$$LRT_0 = \frac{|\widehat{S}_n|^{n/2}}{\left(\operatorname{tr}(\widehat{S}_n)/p\right)^{pn/2}}.$$

This test statistic is proved to exist only when $n \geq p - 1$. When $p \geq n$, LRT is no longer valid because with probability one the determinant $|\widehat{S}_n| = 0$.

The following proposition describes the limiting distribution of the test statistics (Muirhead 1982 (Theorem 8.3.7)).

Proposition 2.4.1 *Under H_0, the limiting distribution of the test statistic is*

$$-2\rho \ln LRT_0 \xrightarrow{d} \chi^2_{(df)},$$

where ρ is a finite sample correction factor that is equal to $\rho = 1 - \dfrac{2p^2 + p + 2}{6pn}$, and the degrees of freedom are $df = \dfrac{1}{2}p(p+1) - 1$.

Thus, in the traditional setup with fixed p, under H_0 for $n \to \infty$ the test statistic $-2\rho \ln LRT_0$ is asymptotically chi-square (χ^2) distributed with df degrees of freedom.

If the null hypothesis is rejected at $\alpha\%$ significance level, this means that the population covariance matrix Σ being equal to the identity, $\Sigma = I$, or its proportional, $\Sigma = \tau^* I$, is not supported by the data, and hence, the true covariance matrix might have a more sophisticated structure.

When addressing the possible population covariance structures, one should differentiate among the following three cases. In the first case, the covariance matrix is equal to an identity matrix $\Sigma = I$ or $\Sigma = \tau^* I$, $\tau^* > 0$, and thus, there is no possibility for a dimension reduction. This can be detected by the sphericity test. In the second case, the covariance matrix is such that after the diagonal transformation all eigenvalues are non-unit and distinct: $\Sigma = \operatorname{diag}(\tau_1, \ldots, \tau_p)$, and $\tau_1 > \ldots > \tau_p > 1$.

The third case is a combination of the two former cases, where it is assumed that the k smallest population eigenvalues are positive and equal to each other, while the largest $p - k + 1$ eigenvalues are non-unit and distinct: $\Sigma = \operatorname{diag}(\tau_1, \ldots, \tau_{p-k+1}, \tau^*, \ldots, \tau^*)$. This third case is of great interest in various fields of study, as it is often associated with the existence of a few factors underlying the data structure, which, in turn, is a compelling model of reality for many applications. The part of a covariance matrix corresponding to distinct eigenvalues is often related to the factors or "signals," while another part with equal eigenvalues corresponds to the "noise." Furthermore, this structure of the covariance matrix implies a reduced rank of order $p - k + 1$.

An extension of the results for the sample eigenvalues and eigenvectors to the case when the population covariance matrix is of the combined form $\Sigma = \text{diag}(\tau_1, \ldots, \tau_{p-k+1}, \tau^*, \ldots, \tau^*)$ is given in Anderson (1963) (Section 13.5.2). The sample eigenvalues and eigenvectors corresponding to signal part, i.e., $\tau_1 > \tau_2 > \ldots > \tau_{p-k+1} > \tau^*$, are asymptotically normally distributed according to Propositions 2.2.3–2.3.1. However, the sample counterparts associated with the noise part, $\tau_{p-k} = \ldots = \tau_p = \tau^*$, are distributed as the eigenvalues of the Wigner matrix (Proposition 2.2.4), and the corresponding sample eigenvectors are conditional Haar distributed (Proposition 2.3.2).

In order to test the reduced rank of the population covariance matrix and, hence, for the presence of factors, the partial sphericity test for the null hypothesis $H_0 : \tau_{p-k} = \ldots = \tau_p = \tau^*$ is traditionally employed in multivariate statistics.

As mentioned earlier, if the sphericity test is rejected, there is still a possibility that the $p - k$ smallest population eigenvalues are equal, and in that case, most of the variation in data is explained by the first k principal directions, or PCs.

The partial sphericity test is performed sequentially:

$$H_k : \tau_{k+1} = \ldots = \tau_p = \tau^* \text{ vs. } H_1 : \tau_{k+2} = \ldots = \tau_p = \tau^*,$$

for all $k \in \{0, \ldots, p-2\}$. Hence, for $k = 0$, the sphericity test defined earlier is nested into the partial sphericity test.

Furthermore, the null hypothesis H_k is equivalent to the null hypothesis that the population covariance matrix has a reduced rank, and hence, the so-called factor structure $\Sigma = T + \tau^* I$, where $T = \text{diag}(\tau_1, \ldots, \tau_k, 0, \ldots, 0)$, is positive-semidefinite matrix of rank k. The latter representation implies as well that the population covariance matrix under H_k has k spiked eigenvalues according to the definition introduced by Johnstone (2001). Here it should be stressed that we assume multiplicity[1] 1 for the population eigenvalues.

The test statistic analogous to LRT_0 is given as (Anderson 1963, Section 11.7.3):

$$LRT_k = \frac{\prod_{i=k+1}^{p} \lambda_{n,i}}{\left(\sum_{i=k+1}^{p} \lambda_{n,i}/(p-k)\right)^{(p-k)}}.$$

This test statistic is defined only if $n \geq p - k - 1$.

[1] The algebraic multiplicity of an eigenvalue τ is the power m of the term $(x - \tau)^m$ in the characteristic polynomial $p_A(x) = \det(xI - A)$, where τ is the eigenvalue of a matrix A. The geometric interpretation of multiplicity is the number of linearly independent eigenvectors associated with an eigenvalue.

The following proposition is from Muirhead (1982) (Theorem 9.6.2). The original result is from Lawley (1956), with the finite sample corrections provided by James (1969).

Proposition 2.4.2 *Under H_k, the limiting distribution of the test statistic is*

$$-\rho \ln LRT_k \xrightarrow{d} \chi^2_{(df)},$$

where a finite sample correction factor ρ is given by

$$\rho = (n-k) - \frac{2(p-k)^2 + (p-k) + 2}{6(p-k)} + \sum_{i=1}^{k} \frac{\bar{\lambda}_k^2}{(\lambda_{n,i} - \bar{\lambda}_k)^2},$$

with $\bar{\lambda}_k = \dfrac{1}{p-k} \sum_{i=k+1}^{p} \lambda_{n,i}$, and the degrees of freedom are

$$df = \frac{1}{2}(p-k)(p-k+1) - 1.$$

Thus, in the traditional setup with fixed p, under H_k for $n \to \infty$ the test statistic $\rho \ln LRT_k$ is asymptotically chi-square (χ^2) distributed with df degrees of freedom.

However, in the high-dimensional setup ($p, n \to \infty$ and $c_n = p/n \to c \in (0, \infty)$), these tests are no longer asymptotically χ^2 distributed. Moreover, for both tests, the probability of wrongly rejecting the null hypothesis, the nominal size α, tends to 1 (see, e.g., Jiang & Yang 2013), while the power of the partial sphericity test deteriorates and tends to α as c gets close to 1 (Forzani et al. 2017).

In the next chapter, we demonstrate the deficiencies of the standard estimators and statistical tools in the high-dimensional setting.

References

Akemann, G., Baik, J., & Di Francesco, P. (2011). *The Oxford handbook of random matrix theory*. Oxford University Press.

Anderson, T. W. (1963). Asymptotic theory for principal component analysis. *Annals of Mathematical Statistics, 34*(1), 122–148.

Forzani, L., Gieco, A., & Tolmasky, C. (2017). Likelihood ratio test for partial sphericity in high and ultra-high dimensions. *Journal of Multivariate Analysis, 159*, 18–38.

Friedman, J. H. (1989). Regularized discriminant analysis. *Journal of the American Statistical Association, 84*(405), 165–175.

James, A. (1969). Tests of equality of latent roots of the covariance matrix. In P.R. Krishnaiah (Ed.), *Multivariate analysis* (Vol. II, pp. 205–218). Academic Press.

Jiang, T., & Yang, F. (2013). Central limit theorems for classical likelihood ratio tests for high-dimensional normal distributions. *The Annals of Statistics, 41*(4), 2029–2074.

Johnstone, I. M. (2001). On the distribution of the largest eigenvalue in principal components analysis. *Annals of Statistics, 29*(2), 295–327.

Kollo, T., & von Rosen, D. (1995). Approximating by the Wishart distribution. *Annals of the Institute of Statistical Mathematics, 47*(4), 767–783.

Lawley, D. N. (1956). Tests of significance for the latent roots of covariance and correlation matrices. *Biometrika, 43*(1–2), 128–136.

Lawley, D. N. (2008). On the asymptotic behavior of the sample estimates of eigenvalues and eigenvectors of covariance matrices. *IEEE Transactions on Signal Processing, 56*(11), 5353–5368.

Lawley, D. N. (2015). Spectrum estimation: A unified framework for covariance matrix estimation and PCA in large dimensions. *Journal of Multivariate Analysis, 139*, 360–384.

Muirhead, R. J. (1978). Latent roots and matrix variates: A review of some asymptotic results. *The Annals of Statistics, 6*(1), 5–33.

Muirhead, R. J. (1982). *Aspects of multivariate statistical theory*. Wiley series in probability and statistics.

Paul, D., & Aue, A. (2014). Random matrix theory in statistics: A review. *Journal of Statistical Planning and Inference, 150*, 1–29.

Rigolett, P. (2015). *High-dimensional statistics*. Lecture notes. MIT

Roy, S. (1953). On a heuristic method of test construction and its use in multivariate analysis. *The Annals of Mathematical Statistics, 24*(2), 220–238.

Srivastava, M. S., & Khatri, C. G. (1979). *An introduction to multivariate statistics*. New York: North-Holland.

Wishart, J. (1928). The generalised product moment distribution in samples from a normal multivariate population. *Biometrika, 20A*(1–2), 32–52.

Chapter 3
Finite Sample Performance of Traditional Estimators

Abstract We demonstrate that well-known multivariate statistical techniques perform poorly and become misleading, when the data dimension p is comparable in magnitude to or larger than the sample size n.

Keywords "large p · large n" setting · Finite sample · Inconsistency

In this chapter, we study the performance of the traditional estimators and multivariate statistical techniques in the setup in which the dimension p is comparable in magnitude to the sample size n. This setting approximates very well the high-dimensional asymptotics framework, $p, n \to \infty$ and $c_n = p/n \to c \in (0, \infty)$, in finite samples.

The primary example we work with in this section is an identity population covariance matrix, as it facilitates discussion about the theoretical properties of traditional estimators under high-dimensional asymptotics. The results in Appendices A–B provide additional evidence for an arbitrary population covariance matrix.

This chapter demonstrates the inconsistency of the sample covariance matrix and its eigenvalues and eigenvectors in finite samples by means of Monte Carlo replications. Furthermore, it shows the poor performance of sphericity and partial sphericity tests in the given setup. This, in turn, implies that traditional estimators and tests may yield misleading and unreliable results when working with small sample sizes.

Hence, this chapter complements the previous one and shows the deficiencies of traditional asymptotic approximations when applied in finite samples.

A. Zagidullina, *High-Dimensional Covariance Matrix Estimation*,
SpringerBriefs in Applied Statistics and Econometrics,
https://doi.org/10.1007/978-3-030-80065-9_3

Table 3.1 Sample covariance matrix. Amplification of the estimation error. $n = 100$

	c_n	$\left\|\widehat{S}_n - \Sigma_n\right\|_F^2$	$\left\|\widehat{S}_n - \Sigma_n\right\|_1^2$	$\left\|\widehat{S}_n - \Sigma_n\right\|$
p = 30, n = 100	0.3	9.31	10.76	1.25
p = 60, n = 100	0.6	36.64	40.22	2.00
p = 90, n = 100	0.9	81.96	87.56	2.64

Note: The distribution of random variables comprising the $n \times p$ data matrix Y_n is real Gaussian with mean zero and $\Sigma = I$. The entries in the table are the average values over $R = 10{,}000$ Monte Carlo replications. The subscript n in c_n emphasizes the fact that this is a finite sample ratio and is not the limiting value c

3.1 Sample Covariance Matrix

In finite samples where the dimension p is of the same order as the number of observations n, the traditional covariance matrix estimator performs poorly. This stems from the intrinsic link between the matrices and their eigenvalues that we discussed earlier in Example 2.1.1.

Similar to the SVD the sample covariance matrix has the following eigendecomposition[1] in terms of eigenvalues and eigenvectors: $\widehat{S}_n = U_n \Lambda_n U_n'$, where $U_n = (u_{n,1}, \ldots, u_{n,p})$ is the matrix of sample eigenvectors, and $\Lambda_n = \text{diag}(\lambda_{n,1}, \ldots, \lambda_{n,p})$ is the diagonal matrix of sample eigenvalues, where $\lambda_{n,1} \geq \lambda_{n,2} \cdots \geq \lambda_{n,p}$. Thus, if the sample eigenvalues and eigenvectors do not comply with the asymptotic results in the finite samples (see, e.g., Sect. 2.2.2 for finite sample bias), neither does the sample covariance matrix. Furthermore, if p is larger than n, the sample covariance matrix $\widehat{S}_n$ is rank-deficient and thus is no longer invertible.

Tables 3.1 and 3.2 below demonstrate that the sample covariance matrix $\widehat{S}_n$ is not a consistent estimator of the population counterpart. In the high-dimensional setup, the Frobenius norm, $\|\cdot\|_F^2$, the operator norm, $\|\cdot\|$, and the matrix l_1-norm, $\|\cdot\|_1^2$, of the difference $\widehat{S}_n - \Sigma$ do not converge to zero as the sample size n tends to infinity for a fixed dimension-to-sample ratio c_n.

Table 3.3 on the contrary presents the results under traditional asymptotics, when p is fixed and n increases. The consistency of the sample covariance matrix is apparent in this setting. However, in order to achieve this property we need $n = 10{,}000$ samples, if the dimension is fixed to $p = 30$.

Tables B.1, B.2, and B.3 in Appendix B provide further results in finite samples and under the asymptotic regime for the general population covariance matrix $\Sigma_n = \text{diag}(\tau_{n,1}, \ldots, \tau_{n,p})$. The eigenvalues of this matrix, $\tau_{n,1}, \ldots, \tau_{n,p}$, are

[1] In fact, eigendecomposition is a special case of SVD applied to symmetric positive definite matrices. SVD is, in turn, valid for any rectangular matrices Jolliffe (2002).

Table 3.2 Sample covariance matrix. Amplification of the estimation error. $n = 1000$

	c_n	$\left\|\widehat{S}_n - \Sigma_n\right\|_F^2$	$\left\|\widehat{S}_n - \Sigma_n\right\|_1^2$	$\left\|\widehat{S}_n - \Sigma_n\right\|$
p = 300, n = 1000	0.3	90.29	76.04	1.37
p = 600, n = 1000	0.6	360.61	290.47	2.12
p = 900, n = 1000	0.9	810.94	641.11	2.77

Note: The distribution of random variables comprising the $n \times p$ data matrix Y_n is real Gaussian with mean zero and $\Sigma = I$. The entries in the table are the average values over $R = 10{,}000$ Monte Carlo replications

Table 3.3 Sample covariance matrix. Consistency property

	c_n	$\left\|\widehat{S}_n - \Sigma\right\|_F^2$	$\left\|\widehat{S}_n - \Sigma\right\|_1^2$	$\left\|\widehat{S}_n - \Sigma\right\|$
p = 30, n = 100	0.3	9.30	10.76	1.25
p = 30, n = 500	0.06	1.86	2.01	0.50
p = 30, n = 1000	0.03	0.93	1.00	0.35
p = 30, n = 10000	0.003	0.09	0.10	0.11

Note: The distribution of random variables comprising the $n \times p$ data matrix Y_n is real Gaussian with mean zero and $\Sigma = I$. The entries in the table are the average values over $R = 10{,}000$ Monte Carlo replications

constructed following the design of Ledoit and Wolf (2015), as it resembles very well the eigenvalue structure observed for financial data, i.e., it is right-skewed and with an exponential type of decay.

Moreover, the numbers in the tables above corresponding to the operator norm are in line with the theoretical finite sample result derived in Vershynin (2018) (see, e.g., Section 4.7). The author derives a probability bound for the expected operator norm of the difference $\widehat{S}_n - \Sigma$ and shows that the covariance matrix estimation is guaranteed to have a small relative error ε, defined as $E(\|\widehat{S}_n - \Sigma\|) \leq \varepsilon\|\Sigma\|$, if the dimensionality p is proportional to the sample size n, such that the following relation $p \sim n \cdot \varepsilon^2$ holds true. Indeed, for $p = 30$ and $n = 10000$, the expected relative error is predicted to be around 0.05.

Another problem arising in finite samples is associated with the inverse of the covariance matrix estimator. The inversion of the sample covariance matrix $\widehat{S}_n$ amplifies the estimation noise accumulated in the entries of the matrix when p is comparable to n. This is illustrated in Table 3.4, where the Frobenius norm for the difference $\widehat{S}_n - \Sigma$ and the condition number $\lambda_{max}/\lambda_{min}$[2] of $\widehat{S}_n$ are presented. As

[2] The condition number $\lambda_{max}/\lambda_{min}(A)$ of a matrix A measures how much the solution or output vector x of an equation $Ax = b$ changes for a small change in the input vector b. So in the context of covariance matrix estimation, this is a measure of inaccuracy for the estimator.

Table 3.4 Inverse of sample covariance matrix. Amplification of the estimation error. $n = 100$

	c_n	$\left\|\widehat{S}_n^{-1} - \Sigma_n^{-1}\right\|_F$	$\lambda_{max}/\lambda_{min}$
p = 30, n = 100	0.3	5.87	9.80
p = 60, n = 100	0.6	27.80	50.64
p = 90, n = 100	0.9	352.75	1021.54

Note: The distribution of random variables comprising the $n \times p$ data matrix Y_n is real Gaussian with mean zero and $\Sigma = I$. The entries in the table are the average values over $R = 10{,}000$ Monte Carlo replications

Table 3.5 Inverse of sample covariance matrix. Reduction of estimation noise

	c_n	$\left\|\widehat{S}_n^{-1} - \Sigma^{-1}\right\|_F$	$\lambda_{max}/\lambda_{min}$
p = 30, n = 100	0.3	5.88	9.82
p = 30, n = 500	0.06	1.54	2.53
p = 30, n = 1000	0.03	1.03	1.92
p = 30, n = 10000	0.003	0.31	1.23

Note: The distribution of random variables comprising the $n \times p$ data matrix Y_n is real Gaussian with mean zero and $\Sigma = I$. The entries in the table are the average values over $R = 10{,}000$ Monte Carlo replications

the dimension-to-sample ratio c_n grows, the inverse of the sample covariance matrix gets more inaccurate (the Frobenius norm of $|\widehat{S}_n^{-1} - \Sigma_n^{-1}|$ increases) and instable (the condition number inflates).

In contrast to the high-dimensional setup, under traditional asymptotics, when $c_n \to \infty$, the estimation noise for the inverse of the sample covariance matrix tends to zero (see Table 3.5).

The result for the general non-identity population covariance matrix in Table B.4 in Appendix B shows that the amplification of the estimation error is even more severe in that case. However, as the number of observations n increases for a fixed p the noise in the inverse of sample covariance matrix, on contrary, becomes negligible (see Table B.5).

Furthermore, as can be expected in finite samples, the empirical density function of the estimators $\widehat{\sigma}_{ij}$s does not match the limiting distribution suggested by the traditional asymptotics result. This is also indicated by the Jarque–Bera test as the normality hypothesis is rejected at 1% significance level (see, e.g., Fig. 3.1).

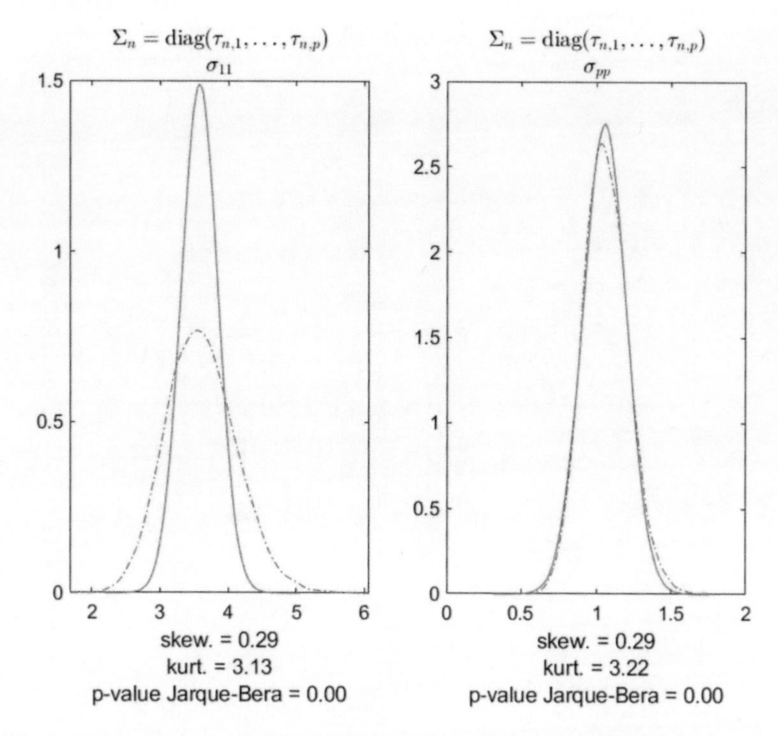

Fig. 3.1 Limiting distribution (solid line) and empirical density (dashed line) of $\widehat{\sigma}_{11}$ and $\widehat{\sigma}_{pp}$. The density of the estimators $\widehat{\sigma}_{11}$ and $\widehat{\sigma}_{pp}$ is estimated non-parametrically with the "Normal" kernel density estimator. The distribution of random variables comprising the $n \times p$ data matrix Y_n is real Gaussian with $\Sigma = \text{diag}(\tau_1, \ldots, \tau_p)$. The i-th population eigenvalue is equal to $\tau_i = H^{-1}((i - 0.5)/p)$, $i = 1, \ldots, p$, where the limiting spectral distribution H is given by the distribution of $1 + 10Z$, and $Z \sim Beta(1, 10)$. The number of Monte Carlo replications is $R = 10{,}000$. $n = 100$ and $p = 60$

3.2 Sample Eigenvalues

It is well-known that the estimated largest sample eigenvalues are upward-biased, while the smallest ones are downward-biased. This bias gets more pronounced in finite samples as the number of observations reduces, or in other words, as the dimension-to-sample ratio c_n increases (see, e.g., Friedman 1989). Furthermore, sample eigenvalues $\lambda_{n,i}$ also tend to be more spread out than the population eigenvalues τ_i. This effect is strongest for the case in which all population eigenvalues τ_i are the same (e.g., are unit-valued) (Johnstone 2001). These phenomena are illustrated in Fig. 3.2 for the case of an identity population covariance matrix, $\Sigma = I$. However, the same issues arise as well in the case of an arbitrary population covariance structure.

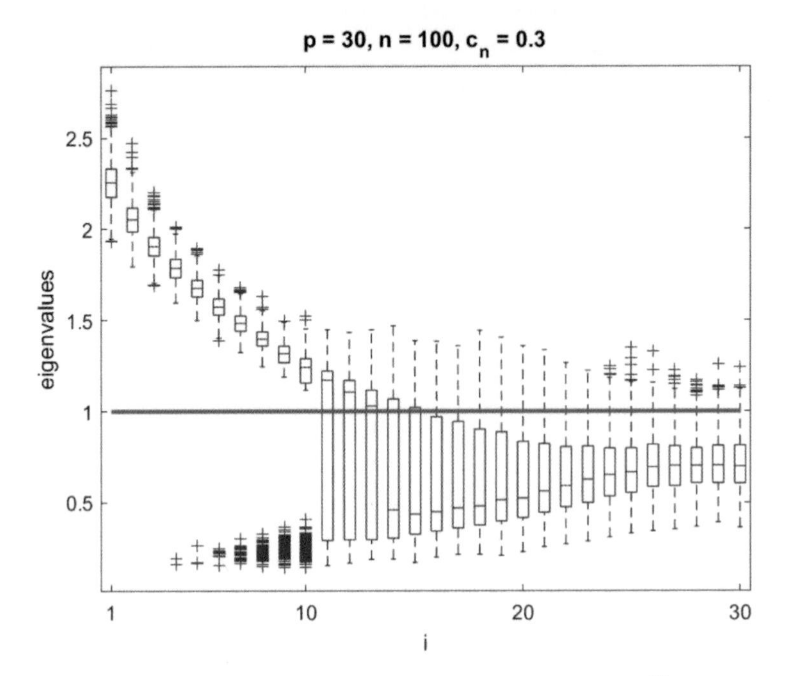

Fig. 3.2 The boxplots for the eigenvalues of the sample covariance matrix $\widehat{S}_n$ over $R = 1000$ Monte Carlo replications. The blue thick line indicates the unit population eigenvalues. The distribution of random variables comprising the $n \times p$ data matrix Y_n is real Gaussian with mean zero and $\Sigma = I$

Such finite sample properties of the eigenvalue estimators have severe implications, for instance, in financial applications. And in particular, the downward bias of the smallest sample eigenvalue leads to the unreliable conclusions in the Markowitz portfolio optimization problem. In this context, the population counterpart (the smallest population eigenvalue) measures the risk of investing in a portfolio of stocks (see, e.g., Campbell et al. 1997, Chapter 5). As a consequence, a significant under-estimation of this risk, or in other words too optimistic assessment of potential losses, can result in wrong decisions regarding the investment strategy.

The problem of potential significant risk under-estimation stemming from a bias of the smallest sample eigenvalue is demonstrated in Table 3.6 using the Monte Carlo replications. For that, we consider an arbitrary population covariance matrix of the form $\Sigma = VTV'$, where $V = (v_1, \ldots, v_p)$ is the matrix of population eigenvectors, and $T = \mathrm{diag}(\tau_1, \ldots, \tau_p)$ is the diagonal matrix containing population eigenvalues $\tau_1 > \ldots > \tau_p > 1$, constructed such that $\tau_i = 1 + 1/i$ for $i = 1, \ldots, p$. Thus, the resultant population covariance matrix Σ is of full rank p. The corresponding sample covariance matrix $\widehat{S}_n$ has positive and distinct eigenvalues in each realization of the Monte Carlo replications, if the dimension-to-sample ratio is $c_n = 0.8$ ($p = 200$ and $n = 250$). As a consequence, it can be considered as the full-rank matrix. However, the mean of the smallest sample

Table 3.6 Inconsistency of sample eigenvalues in the high-dimensional setup, $c_n = 0.8$, $p = 200$, $n = 250$

τ_1	$\lambda_{n,1}$	τ_2	$\lambda_{n,2}$	τ_p	$\lambda_{n,p}$
2.00	3.71	1.501	3.54	1.01	0.01
	(0.1019)		(0.0717)		(0.0019)
	$\cos\phi(u_{n,1}, v_1)$		$\cos\phi(u_{n,2}, v_2)$		$\cos\phi(u_{n,p}, v_p)$
	0.34		0.11		0.06
	(0.1680)		(0.0829)		(0.0417)

Note: The distribution of random variables comprising the $n \times p$ data matrix Y_n is real Gaussian with mean zero and covariance matrix Σ. The true covariance matrix is computed according to the eigendecomposition $\Sigma = V \cdot T \cdot V'$, where V denotes the $(p \times p)$ orthogonal matrix of eigenvectors, and $T = \text{diag}(\tau_1, \ldots, \tau_p)$ is a diagonal matrix containing the eigenvalues $\tau_i = \frac{i+1}{i}$ (for $i = 1, \ldots, p$). The numbers indicate the mean over the $R = 1000$ Monte Carlo replications, and the numbers in parentheses indicate the corresponding standard errors. v_i is a true i-th eigenvector, and $u_{n,i}$ denotes the estimated i-th sample eigenvector

eigenvalue $\lambda_{n,p}$ is equal to 0.01, which is very close to zero, and thus, implies the rank deficiency of the sample covariance matrix. Furthermore, it is 100 times smaller than the population counterpart $\tau_p = 1.01$ and, hence, is heavily underestimated. In the portfolio optimization context this means that the potential risk of the asset allocation is substantially understated.

Tables B.6 and B.7 in Appendix B present additional results for this example when the dimension-to-sample ratio c_n takes different values. One can observe the following pattern: as the dimension-to-sample ratio c_n decreases from 0.8 to 0.5 and then further to 0.2, the bias of the sample eigenvalues also gradually decreases as we get closer to the traditional asymptotics regime where $c_n \to 0$ as $n \to \infty$.

Furthermore, in the case when $p > n$, the sample covariance matrix $\widehat{S}_n$ has the "spurious" zero eigenvalues $\lambda_{n,i}$ for $i = n+1, \ldots, p$ that actually correspond to the positive population counterparts τ_i and appear only due to the rank deficiency. Such spurious close to zero realizations of sample eigenvalues can be observed already for the case of $p = 100$, $n = 100$, and $c_n = 1$ (see, e.g., Fig. A.1 in Appendix A).

3.3 Sample Eigenvectors

The performance of respective eigenvector estimators as well degrades in finite samples when the dimension p is comparable to the sample size n.

The simulation results given in Table 3.6 (and in Tables B.6 and B.7, Appendix B) demonstrate the finite sample properties of the eigenvector estimators in the case of a general population covariance matrix, $\Sigma = VTV'$, which is of full rank. The finite sample constellations of p and n in the tables approximate the high-dimensional

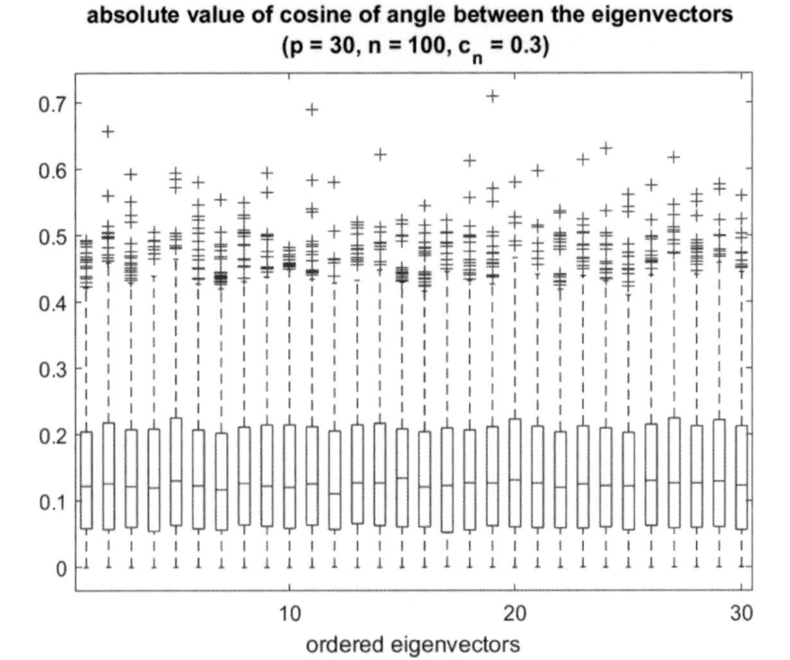

Fig. 3.3 The boxplots of $\cos\phi(u_{n,i}, v_i)$, the cosine of the angle between the sample eigenvector and the true one over $R = 1000$ Monte Carlo replications. The distribution of random variables comprising the $n \times p$ data matrix Y_n is real Gaussian with mean zero and $\Sigma = I$

asymptotics ($p, n \to \infty$, $c_n = p/n \to c > 0$). From the tables one may observe that as the dimension-to-sample ratio c_n increases to 1, the mean of the angle ϕ estimated between the population v_i and sample $u_{n,i}$ eigenvectors grows to 90° (the corresponding average cosine of the angle, $\cos\phi$, gets closer to 0). This, in turn, implies that the sample eigenvectors move farther away from the true ones in an orthogonal direction when the sample size n decreases for a given dimension p.

In addition, Fig. 3.3 presents the estimation results for the cosine of an angle between the true and estimated eigenvectors, for the case in which the population covariance matrix is identity, $\Sigma = I$.

According to the boxplots, the mean of $\cos\phi$ is about 0.12 for all p sample eigenvectors. Thus, the angle ϕ between the sample and the population eigenvectors on average is about 83.1°, meaning that the vectors are almost orthogonal to each other in this case as well. This empirical outcome, further, describes very well the theoretical finding that the sample eigenvectors are randomly distributed on the unit sphere in the case of the identity population covariance matrix (see Sect. 2.3, Propositions 2.3.2–2.3.4). The same result for the case $p = 100$, $n = 100$ and $c_n = 1$ is presented in Fig. A.2 (Appendix A).

The above results are essentially the finite sample approximations to the inconsistency of eigenvector estimators established under the high-dimensional asymptotics for the case when the population covariance matrix is of full rank. However, the same property is valid when the true covariance matrix has the factor structure, and hence, the reduced rank. For example, Johnstone and Lu (2004) prove that eigenvectors are estimated inconsistently as well in the case of a rank one population covariance matrix (the special case of only one factor underlying the true covariance structure). Further details of this proof are presented in Sect. 4.4. Moreover, in a general case with more than one influential factor, the sample eigenvectors are also inconsistent.

The fact that space spanned by the sample eigenvectors $u_{n,i}$ does not coincide with the true eigenvector space implies the inconsistency of sample PCs. Thus, the PCA in its standard form cannot be applied in the high-dimensional setting as a dimension-reduction technique or for the estimation of factor models, since the factors and respective factor loadings would be estimated inconsistently (see, e.g., Harding 2007).

Further evidence regarding unreliable PCA estimation results, though in the context of scree plots and choice of the number of PCs, is provided in the next section.

3.4 PCA Applications

The number of PCs in the PCA is often chosen by practitioners based on scree plots, in which one seeks for the "elbow" or "kink" indicating a sudden drop in the ordered eigenvalues. This kink separates the two clusters of sample eigenvalues belonging to the signal and noise subsets.

The following graphs illustrate that in finite samples where the dimension p is comparable to the sample size n, scree plots are misleading when used to choose the number of principal directions. From ordered sample eigenvalues we cannot learn anything about the true population structure of the eigenvalues.

The first graph presents the scree plot based on the true population structure, where all population eigenvalues are unit-valued, i.e., $\tau_i = 1$ for $i = 1, \ldots, p$. The second graph presents the case when one half of the population eigenvalues are equal to 1, and the other half are equal to 2, i.e., $\tau_i = 1$ for $i = 1, \ldots, [p/2]$, and $\tau_i = 2$ for $i = [p/2] + 1, \ldots, p$ (see Fig. 3.4).

In both situations, it is not possible to detect the true distribution of the population eigenvalues just by looking at the scree plots (see Fig. 3.5).

Figures A.3 and A.4 in Appendix A present the same results for the case $p = 100$, $n = 100$, and $c_n = 1$.

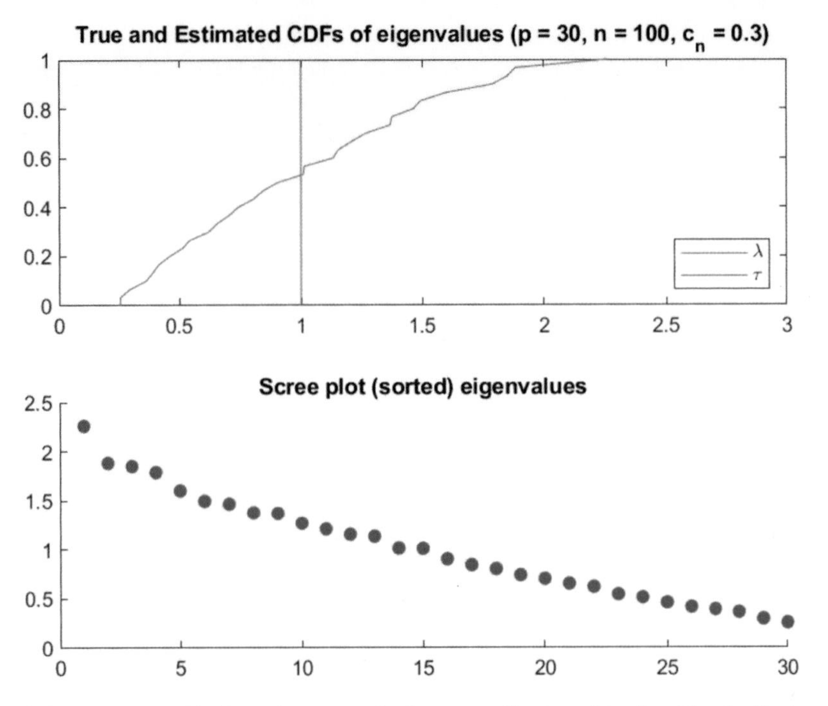

Fig. 3.4 Case $\Sigma = I$. The figure is generated using one realization of the data. The distribution of random variables comprising the $n \times p$ data matrix Y_n is real Gaussian with mean zero and $\Sigma = I$. The c.d.f. of the true eigenvalues is a point mass at 1, δ_1. Based on Figure 1 of El Karoui (2008)

3.5 Sphericity Test

The performance of the sphericity test in finite samples corresponding to the high-dimensional setup is illustrated based on a Monte Carlo study with the true covariance matrix $\Sigma = I$ under the null hypothesis H_0 and $\Sigma = \text{diag}(1.65, \ldots, 1.65, 1, \ldots, 1)$ under the alternative hypothesis H_1.

The empirical size of the sphericity test, $\widehat{\mathbb{P}}\{H_0 \text{ is rejected}|H_0 \text{ is true}\} = \widehat{\alpha}$, and the power, $\widehat{\mathbb{P}}\{H_0 \text{ is rejected}|H_1 \text{ is true}\}$, in such a setting are presented in Table 3.7.

Based on the results given in Table 3.7, we can conclude that the empirical size of the test $\widehat{\alpha}$ tends to 1 with c_n growing to 1, while the nominal size of the test α is supposed to be 0.05. Thus, the results of the sphericity test are not reliable in the high-dimensional setting when p is comparable to n, as the test tends to reject the true hypothesis more often than in 5% of the cases.

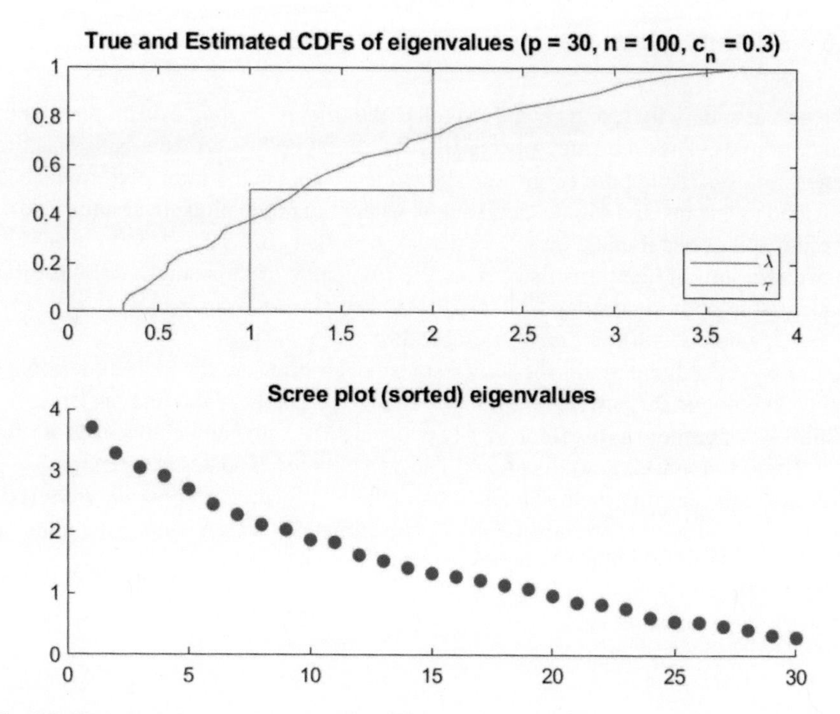

Fig. 3.5 Case $\Sigma = I_{(p/2)} + 2 \cdot I_{(p/2)}$. The figure is generated using one realization of the data. The distribution of random variables comprising the $n \times p$ data matrix Y_n is real Gaussian with mean zero and $\Sigma = I_{(p/2)} + 2 \cdot I_{(p/2)}$. The c.d.f. of the true eigenvalues are point masses at 1 and at 2, i.e., $0.5\delta_1 + 0.5\delta_2$. Based on Figure 3 of El Karoui (2008)

Table 3.7 Sphericity test results; empirical vs. nominal size

	c_n	Size $\widehat{\alpha}$	Power
n = 100, p = 5	0.05	0.0521	0.6505
n = 100, p = 30	0.3	0.0620	0.8384
n = 100, p = 60	0.6	0.3276	0.9633
n = 100, p = 90	0.9	1.0000	1.0000

Note: The empirical size $\widehat{\alpha}$ is estimated based on $R = 10000$ replications with $\Sigma = I$. The empirical power is estimated under the alternative hypothesis with $\Sigma = \text{diag}(1.65, \ldots, 1.65, 1, \ldots, 1)$, where the number 1.65 on the diagonal is equal to $[p/2]$. Based on Table 1 of Jiang and Yang (2013)

3.6 Partial Sphericity Test

The partial sphericity test also tends to reject the true null hypothesis more often in finite samples. The frequency of rejecting the H_0 hypothesis, $\widehat{\alpha}$, tends to 1 as the dimension-to-sample ratio c_n grows. The evidence for such a poor performance is provided by means of a Monte Carlo study, where the population covariance matrix has four spikes or "signals," i.e., $\Sigma = \text{diag}(7, 6, 5, 4, 1, \ldots, 1)$.

We measure the empirical size $\widehat{\alpha}$ as the frequency of choosing $k = 4$ for the sequential hypothesis testing $H_k : \tau_{k+1} = \ldots = \tau_p = \tau^*$ vs. $H_1 : \tau_{k+2} = \ldots = \tau_p = \tau^*$, where $k \in \{0, \ldots, p-2\}$ (Table 3.8).

The power of the partial sphericity test also deteriorates in the high-dimensional setup. The empirical power, $\widehat{\mathbb{P}}\{H_0 \text{ is rejected}|H_1 \text{ is true}\}$, of the test presented in Table 3.9 is measured as the frequency of choosing a certain number of spikes during $R = 10{,}000$ Monte Carlo replications. The true number of spikes is equal to 4.

As we can see from the results in Table 3.9, already with $p = 60$ and $n = 100$, the partial sphericity test tends to either overestimate or underestimate the number of spikes, k, in 45% of the cases.

Table 3.8 Partial sphericity test results; empirical vs. nominal size

	c_n	Size $\widehat{\alpha}$
n = 100, p = 20	0.2	0.0503
n = 100, p = 30	0.3	0.0596
n = 100, p = 60	0.6	0.2811
n = 100, p = 90	0.9	0.9999

Note: The empirical size $\widehat{\alpha}$ is estimated based on $R = 10000$ replications. The distribution of random variables comprising the $n \times p$ data matrix Y_n is real Gaussian with mean zero and Σ, where $\Sigma = \text{diag}(7, 6, 5, 4, 1, \ldots, 1)$ is a spiked population covariance model with the number of spikes $k = 4$. The design is according to Forzani et al. (2017)

Table 3.9 Partial sphericity test results; empirical power

	c_n	$k = 1$	$k = 2$	$k = 3$	$\boldsymbol{k = 4}$	$k = 5$	$k = 6$	$k = 7$
n = 100, p = 20	0.2	0.00	0.00	0.00	0.96	0.02	0.01	0.00
n = 100, p = 30	0.3	0.00	0.00	0.04	0.91	0.04	0.01	0.00
n = 100, p = 60	0.6	0.00	0.00	0.17	0.55	0.11	0.06	0.04
n = 100, p = 90	0.9	0.00	0.00	0.00	0.00	0.00	0.00	0.00 [a]

Note: The values of k selected for the number of spikes over $R = 10000$ replications. The distribution of random variables comprising the $n \times p$ data matrix Y_n is real Gaussian with mean zero and Σ, where $\Sigma = \text{diag}(7, 6, 5, 4, 1, \ldots, 1)$ is a spiked population covariance model with the number of spikes $k = 4$. The design is according to Forzani et al. (2017)

[a] In the case $n = 100$, $p = 90$, the first non-zero frequency for the number of spikes appears for $k = 14$. It is chosen in 1% of the cases

References

Campbell, J. Y., Lo, A. W., & MacKinlay, A. (1997). *The econometrics of financial markets*. Princeton University Press.

El Karoui, N. (2008). Spectrum estimation for large dimensional covariance matrices using random matrix theory. *Annals of Statistics, 36*(6), 2757–2790.

Forzani, L., Gieco, A., & Tolmasky, C. (2017). Likelihood ratio test for partial sphericity in high and ultra-high dimensions. *Journal of Multivariate Analysis, 159*, 18–38.

Friedman, J. H. (1989). Regularized discriminant analysis. *Journal of the American Statistical Association, 84*(405), 165–175.

Harding, M. C. (2007). *Structural estimation of high-dimensional factor models: Uncovering the effect of global factors on the US economy*. Technical Report. MIT, Dept. of Economics

Jiang, T., & Yang, F. (2013). Central limit theorems for classical likelihood ratio tests for high-dimensional normal distributions. *The Annals of Statistics, 41*(4), 2029–2074.

Johnstone, I. M. (2001). On the distribution of the largest eigenvalue in principal components analysis. *Annals of Statistics, 29*(2), 295–327.

Johnstone, I. M., & Lu, A. Y. (2004). *Sparse principal components analysis*. Technical Report. Stanford University, Dept. of Statistics. Available: https://arxiv.org/abs/0901.4392

Jolliffe, I. T. (2002). *Principal component analysis* (2nd ed.). New York: Springer.

Jolliffe, I. T. (2015). Spectrum estimation: A unified framework for covariance matrix estimation and PCA in large dimensions. *Journal of Multivariate Analysis, 139*, 360–384.

Jolliffe, I. T. (2018). *High-dimensional probability: An introduction with applications in data science*. Cambridge series in statistical and probabilistic mathematics. Cambridge University Press.

Chapter 4
Traditional Estimators and High-Dimensional Asymptotics

Abstract We introduce and describe various classical and modern theoretical results developed within the random matrix theory domain which are related to the covariance matrix estimation, as well as to the factor structure inference in high-dimensional data.

Keywords Random matrix theory (RMT) · Marčenko–Pastur distribution · Tracy–Widom law · Spiked covariance model · Principal component analysis (PCA) · Phase transition phenomena · Sphericity tests

This chapter introduces the reader to the theoretical properties of traditional estimators and multivariate testing procedures under high-dimensional asymptotics when $p,\ n \to \infty$, such that their dimension-to-sample size ratio converges to the limiting value, called the concentration ratio (i.e., $c_n = p/n \to c > 0$).

Unlike in the traditional "fixed p, large n" setup, under high-dimensional asymptotics, two types of limiting behavior are distinguished for the sample eigenvalues and their respective eigenvectors, referred to as the null and the non-null cases in the literature. In this introductory section, we briefly discuss the major findings related to both cases under high-dimensional asymptotics and compare these findings to the known results from the traditional setup. Thus, we provide a short overview of this chapter.

Under the high-dimensional asymptotics, in the null case, the extreme sample eigenvalues (the largest and the smallest) converge in the limit to edges of a distribution that describes the behavior of a majority of the sample eigenvalues, or the bulk. This limiting distribution of sample eigenvalues is linked to the population values through the concentration ratio according to the Marčenko–Pastur equation (see Marčenko & Pastur 1967). This is one of the fundamental results derived for the bulk in the null case. The other major finding, in this case, is the fluctuation result for the extreme sample eigenvalues. Under the high-dimensional asymptotics, the largest sample eigenvalue has the Tracy–Widom distribution (see Tracy & Widom

A. Zagidullina, *High-Dimensional Covariance Matrix Estimation*,
SpringerBriefs in Applied Statistics and Econometrics,
https://doi.org/10.1007/978-3-030-80065-9_4

1996 for $\Sigma = I$; Bao et al. 2015 for a diagonal nonidentity Σ[1]), while the smallest has the reflected Tracy–Widom distribution (see Ma 2012 for $\Sigma = I$).

In the null case, the key findings under high-dimensional asymptotics differ strikingly from the analogous results under standard asymptotics. For example, for unit population eigenvalues, the Marčenko–Pastur equation implies that, in the limit, the sample eigenvalues are distributed according to the Marčenko–Pastur law (in this particular case, the explicit form of the density function is available). In contrast, under the standard asymptotics, the sample eigenvalues behave in the limit jointly as the eigenvalues of a Wigner matrix. Furthermore, in the traditional "fixed p, large n" setup, it is hard to obtain the marginal fluctuation results for the extreme sample eigenvalues. However, under the high-dimensional asymptotics, this problem is resolved. Moreover, for non-unit population values, sample eigenvalues have asymptotic Normal distribution under standard asymptotics, and this distribution applies both to the extreme sample eigenvalues and to the majority (i.e., the bulk). However in the high-dimensional setup, the limiting behavior of the bulk is governed by the generalized Marčenko–Pastur equation, and the fluctuation of the largest sample eigenvalue is described by the generalized Tracy–Widom distribution.

On the contrary, in the non-null case, there are many similarities between the results derived under two asymptotic frameworks. The non-null case is considered when the extreme sample eigenvalues in the limit lie outside the support of the spectral distribution, or the support of the Marčenko–Pastur distribution, and follow laws different from the Tracy–Widom type of fluctuations. The behavior of the extreme sample eigenvalues in the non-null case is implied by the spiked population model for the covariance matrix. This model was first proposed by Johnstone (2001) to explain the phenomenon often observed in real-data applications: the separation of a few extreme sample eigenvalues from the majority. One of the most important implications of the spiked population model is that the bound and fluctuation results of extreme sample eigenvalues are subject to phase transition under the high-dimensional asymptotics (see Baik & Silverstein 2006). The phase transition phenomenon applies when the sample eigenvalues behave in the limit as if the population covariance matrix is an identity given that the non-unit eigenvalues are sufficiently close to one. The measure of closeness to one is defined as a certain threshold that depends on the concentration ratio. When applied to the fluctuation results, the phase transition yields that the sample eigenvalues have Gaussian-type fluctuation in the limit if their spiked population counterparts are larger than the transition threshold. This result is similar to the one under the standard asymptotics. Thus, in the non-null case, one can draw an insightful parallel to the traditional asymptotics and understand how the relative magnitude of dimension p and sample size n affects the consistency property of the sample eigenvalues.

The behavior of respective sample eigenvectors is governed by eigenvalues and, hence, is also different under the null and non-null cases. In the null case sample eigenvectors are Haar distributed under both asymptotic regimes ("fixed p, large

[1] The result holds for the complex sample covariance matrix.

n" and "large p, large n") given that the entries of data are Gaussian (see Anderson 1963). In turn, in the non-null case, the sample sub-eigenvectors of dimension k corresponding to the k spiked population eigenvalues are asymptotically multivariate Gaussian (see Paul 2007). This is reminiscent of the standard result in multivariate analysis. However, the latter result holds for the entire vector of dimension p, and the former finding is valid only for the sub-vector. Moreover, in the "large p, large n" setting, the sample eigenvectors converge in the limit to the eigenvectors that correspond to biased population eigenvalues. Therefore, the sample eigenvectors are inconsistent estimators of the population eigenvectors.

Thus, when comparing the properties of the sample eigenvalues and eigenvectors under two different asymptotic regimes, we see a complete transformation of the results in the null case and an extension of the traditional results to the high-dimensional setting in the non-null case.

The other significant difference between the two asymptotic regimes is concerned with the conditions imposed on the data. The high-dimensional results briefly introduced here generally do not depend on the distribution of data. This phenomenon is called *universality*. On the contrary, the limiting results for spectral statistics and eigenvectors under the standard asymptotics are derived under an assumption of normality imposed on the data.

Since the 1960s, Wigner, Mehta, Dyson, and many others have conjectured that the limiting distributions of spectral statistics are *universal*, in the sense that they are valid beyond the Gaussian setting. In the last decade, much effort has been devoted to demonstrating the universality results for the behavior of both bulk and extreme eigenvalues. It was proved that the limiting behavior of the bulk is universal for both i.i.d. and non-i.i.d. settings (in the latter case, the Lindenberg type condition is imposed on the data) and essentially depends on the first four moments of the data distribution. The universality of the bulk behavior was proved using the Stieltjes transform methodology (see Bai & Silverstein 2010). The universality result of the Tracy–Widom distribution for the largest eigenvalue of the Wishart matrix (for identity population covariance matrix) was first derived by Soshnikov (2002) under an assumption of sub-Gaussian tails, symmetry of the data entries' distribution, and the requirement that concentration ratio is equal to one. The assumption on the symmetry of the distribution and the concentration ratio was subsequently relaxed in Peche and Soshnikov (2008) and Péché (2009), respectively. The universality result of the Tracy–Widom distribution for the smallest eigenvalue of the Wishart matrix complemented the above result for the largest eigenvalue and was established in Feldheim and Sodin (2010). Moreover, the recent "four moments" theorem of Tao and Vu (2012) generalizes the universality discussion in the null case and states that the limiting behavior of spectral statistics is the same as when the entries of the data matrix are i.i.d. Gaussian, given that the first four moments of the data distribution match with those of the standard Gaussian. This result has been further extended for bulk and edge universality in Pillai and Yin (2014) by relaxing the requirements such that only the first two moments of data match that of standard Gaussian given a subexponential tail behavior. Furthermore, the universality results were also shown for the sample eigenvectors in the null case by Bai et al. (2007) and Bai et al. (2011) (further details are provided in Sect. 4.3).

In the non-null case, the universality of phase transition phenomena for the leading eigenvalues in the spiked population model was established by Baik and Silverstein (2006). Bai and Yao (2008) extended this bound result by deriving the Gaussian limiting distributions for the leading sample eigenvalues. Furthermore, Bai and Yao (2012) enhanced both bound and distributional results to the setting of the generalized spiked model, in which the base covariance model is of arbitrary form. Moreover, Féral and Péché (2009) established the Tracy–Widom type universality for the largest eigenvalue of the sample covariance matrix under the spiked population model with a diagonal covariance matrix Σ. In addition, for a general nonidentity diagonal matrix Σ, Bao et al. (2015) and Lee and Schnelli (2016) showed that the limiting distribution of the largest sample eigenvalue corresponding to a non-fundamental spike is also given by the Tracy–Widom distribution, under mild conditions on the data. For the sample eigenvectors under the spiked population model, Bloemendal et al. (2016) established that similar bound results hold as in the Gaussian case (see Paul 2007) for the data distributions with subexponential tails.

However, there are still many open questions related to the universality of the limiting results under high-dimensional asymptotics (the interested reader is referred to Akemann et al. 2011 and Paul and Aue 2014). In general, the distributional assumptions imposed on the data, and hence universality property, are mainly driven by the methodologies chosen to prove specific results. For example, when using the orthogonal polynomial decomposition of the eigenvalues' exact density (e.g., Mehta 1990), it is necessary to impose an assumption on the form of the underlying data distribution (e.g., symmetry and invariance). In contrast, the techniques using Stieltjes transform or the method of moments do not require any density conditions (see Bai & Silverstein 2010).

In this chapter, for ease of exposition, we do not address the universality questions and, in many cases, present the original formulations of results based on the assumption of Gaussian data, as for the Tracy–Widom distribution.

4.1 Sample Covariance Matrix

The properties of high-dimensional sample covariance matrices in finite samples are well approximated under high-dimensional asymptotics. However, if we impose the RMT type of condition on the relative rate of growth for p and n, i.e., $p, n \to \infty$, $c_n = p/n \to c > 0$, consistent estimation of covariance matrices is impossible without any further restrictions on their structure. This problem stems from the fact that the eigenvalues of the sample covariance matrix are no longer consistent estimators for the population values. This is described in more detail in the subsequent sections. Furthermore, even if it is possible to construct the consistent estimators for the population eigenvalues, the simultaneous estimation of the consistent eigenvectors is not yet possible under realistic assumptions.[2] Thus, without any ad hoc structures, such as sparsity or an underlying factor model imposed on the population covariance matrix, the consistent estimation is hard to achieve. The comprehensive reviews on the topic of high-dimensional covariance matrix estimation under this type of ad hoc assumptions are given in Fan et al. (2011) and Fan et al. (2016).

Alternatively, in the absence of the ad hoc type of structures imposed on the population covariance matrix, Ledoit and Wolf (2012, 2015), considered rotation-equivariant shrinkage (or *oracle*) estimator for the covariance matrix under the RMT type of asymptotics. The oracle estimator is based on the eigendecomposition of the sample covariance matrix. It is defined as $\tilde{S}_n^{\text{Oracle}} = U_n \tilde{\Lambda}_n^{\text{Oracle}} U_n'$, where the inconsistent sample eigenvectors U_n are not modified, while the diagonal matrix Λ_n that contains the sample eigenvalues is replaced by matrix $\tilde{\Lambda}_n^{\text{Oracle}}$ with the consistent estimators of the population eigenvalues under high-dimensional asymptotics. This is an application of the general shrinkage principle, going back to Stein (1956) and Haff (1980). This estimator is feasible and consistent under the RMT type of asymptotics, in a sense that it converges to the *oracle* population covariance matrix. This framework has been further extended to linear shrinkage estimators of precision matrices by Wang et al. (2015) and Bodnar et al. (2016), among others. Furthermore, Zhang et al. (2013) generalized this approach to the linear shrinkage estimators of covariance and precision matrices that are not equivariant.

4.2 Sample Eigenvalues

Under high-dimensional asymptotics, two different cases are considered with regard to the behavior of the sample eigenvalues: the bulk, which refers to the properties of the full set of sample eigenvalues $\lambda_{n,1}, \ldots, \lambda_{n,p}$, and the extremes, which address the largest and smallest eigenvalues.

Moreover, there is a distinction between the null and non-null cases. The null case refers to the situation when the extreme eigenvalues are packed together with the bulk or the majority of the eigenvalues. The non-null case describes the situation when the extreme eigenvalues are separated from the bulk and do not follow the limiting laws that are valid for the null case.

[2] Mestre (2008) considers the joint consistent estimation of the eigenvalues and eigenvectors based on the sample counterparts, however, under very specific assumptions. In particular, it is assumed that the multiplicities of the population eigenvalues are known.

4.2.1 Bounds on $\lambda_{n,1}$ and $\lambda_{n,min}$

The bound result for the largest sample eigenvalue in the null case was first provided by Yin et al. (1988), while the same result for the smallest eigenvalue was obtained by Bai and Yin (1993).

Proposition 4.2.1 *Let the entries $\{X_{ij}\}$ of the $n \times p$ data matrix X_n be i.i.d. complex-valued random variables with $E(X_{ij}) = 0$, $E(|X_{ij}|^2) = 1$ and $E(|X_{ij}|^4) < \infty$. Consider the sample covariance matrix $\widehat{S}_n$ and denote its eigenvalues in a decreasing order as $\lambda_{n,1} \geq \lambda_{n,2} \geq \ldots \geq \lambda_{n,p}$. When $p, n \to \infty$ and $c_n = p/n \to c > 0$,*

$$\lambda_{n,1} \to b_c = (1 + \sqrt{c})^2 \; a.s.,$$

$$\lambda_{n,min} \to a_c = (1 - \sqrt{c})^2 \; a.s.,$$

$$\lambda_{n,min} = \begin{cases} \lambda_{n,p}, & \text{for } p \leq n, \\ \lambda_{n,p-n+1}, & \text{for } p > n. \end{cases}$$

A remarkable result is that for $c = 1$, the largest sample eigenvalue $\lambda_{n,1}$ converges almost surely to the value of 4. We can also observe this in Fig. 4.1.

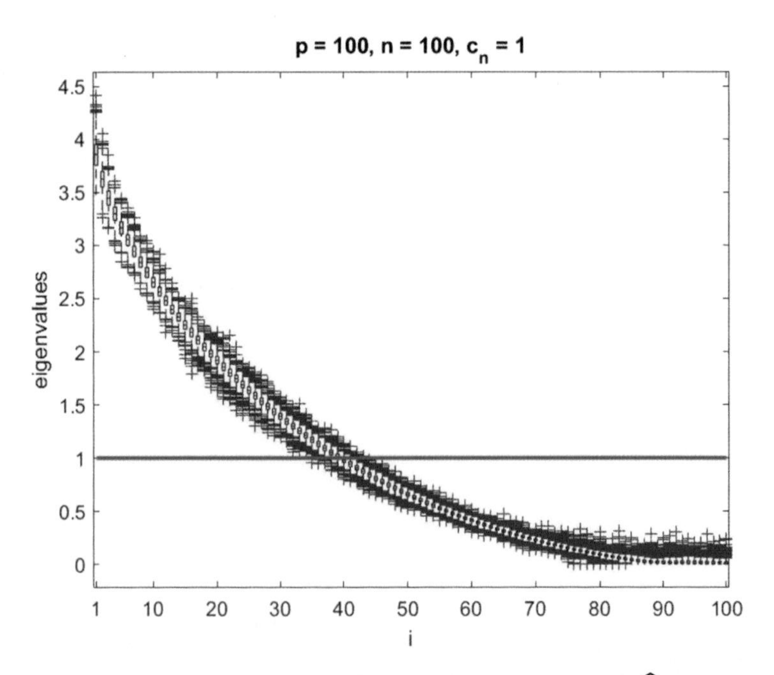

Fig. 4.1 The boxplots for the eigenvalues of the sample covariance matrix $\widehat{S}_n$ over $R = 1000$ Monte Carlo replications. The blue thick line indicates the unit population eigenvalues. The distribution of random variables comprising the $n \times p$ data matrix Y_n is real Gaussian with mean zero and $\Sigma = I$

The average value for the largest sample eigenvalue over $R = 1000$ Monte Carlo replications is around 4. Furthermore, the average value over $R = 1000$ replications for the smallest sample eigenvalue is around 0, which is also predicted by the bound result.

4.2.2 Fluctuation Results for $\lambda_{n,1}$ and $\lambda_{n,min}$

The above-stated bound results for the largest and smallest eigenvalues are remarkable because they explain the relative spread of sample eigenvalues compared to the population ones. However, they do not indicate the variability of the extreme eigenvalues, or in other words, their distribution.

The exact distributional results for the extreme eigenvalues depend on the hypergeometric functions of the matrix argument. These functions converge very slowly, even for small p, thus making it difficult to determine the distribution (see, Sect. 2.2.1).

Under traditional asymptotics, the extreme sample eigenvalues and the ones that correspond to the bulk are asymptotically normally distributed.

In the high-dimensional setting $p,\ n \to \infty$, $c_n = p/n \to c > 0$ the asymptotic distributional results for the largest and smallest eigenvalues change.

The fluctuation result for the largest sample eigenvalue was first introduced by Johnstone (2001). This asymptotic result already provides very good approximations for n and p as small as five.

Proposition 4.2.2 *Let the entries $\{X_{ij}\}$ of the $n \times p$ data matrix X_n be i.i.d. $\mathcal{N}(0, 1)$. Denote as $\lambda_{n,1} > \ldots > \lambda_{n,p}$ the eigenvalues of sample covariance matrix $\widehat{S}_n$. The fluctuation of the largest eigenvalue $\lambda_{n,1}$ is characterized as follows:*

$$\frac{\lambda_{n,1} - \mu_{np}}{\sigma_{np}} \xrightarrow{d} \mathcal{F}_1,$$

$$\mu_{np} = \left(\sqrt{n-1} + \sqrt{p}\right)^2,$$

$$\sigma_{np} = \left(\sqrt{n-1} + \sqrt{p}\right)\left(\frac{1}{\sqrt{n-1}} + \frac{1}{\sqrt{p}}\right)^{1/3},$$

where μ_{np} and σ_{np} are the center and scaling parameters, and $\mathcal{F}_1$ is a Tracy–Widom law of order 1.

The Tracy–Widom law of order 1 is defined in terms of the function that solves the Painléve II differential equations (see, e.g., Tracy and Widom 1996). The distribution function $\mathcal{F}_1$ has no closed-form solution and is evaluated numerically.

It should be noted here that $\mu_{np} \simeq b_c = (1 + \sqrt{c})^2$ for both p and n large, which is in accordance with the bound result in Proposition 4.2.1.

The original result in Johnstone (2001) is stated under the asymptotic regime when $p \to \infty$, $n = n(p) \to \infty$, $\gamma_n = n/p \to \gamma \geq 1$,[3] so that the center and scale parameters grow with p. The largest sample eigenvalue grows with mean of order $\mathcal{O}(p)$, while the standard deviation about that mean is of order $\mathcal{O}(p^{1/3})$.

This implies that the distribution of $\lambda_{n,1}$ is tighter about the mean value compared to the standard central limit theorem (CLT) results, in which the standard deviation is of order $\mathcal{O}(n^{1/2})$.

The center μ_{np} and scaling σ_{np} parameters imply that the difference between the distribution of $\dfrac{\lambda_{n,1} - \mu_{np}}{\sigma_{np}}$ and $\mathcal{F}_1$ is of "second order" $\mathcal{O}((n \wedge p)^{-1/3})$.[4]

Recently, Ma (2012) proposed corrections for the centering $\mu_{np} = \left(\sqrt{n_-} + \sqrt{p_-}\right)^2$ and scaling $\sigma_{np} = \left(\sqrt{n_-} + \sqrt{p_-}\right)\left(\frac{1}{\sqrt{n_-}} + \frac{1}{\sqrt{p_-}}\right)^{1/3}$, with $n_- = n - \frac{1}{2}$, $p_- = p - \frac{1}{2}$ so that the "second order" difference is improved to $\mathcal{O}((n \wedge p)^{-2/3})$.

The result for the smallest sample eigenvalue was shown in Ma (2012) under the condition of the relative rate of convergence for p and n defined as: $p + 1 \leq n$, $p \to \infty$, $n = n(p) \to \infty$, and $\gamma_n = n/p \to \gamma \in (1, \infty)$.

Proposition 4.2.3 *Let the entries $\{X_{ij}\}$ of the $n \times p$ data matrix X_n be i.i.d. $\mathcal{N}(0, 1)$. Denote as $\lambda_{n,1} > \ldots > \lambda_{n,p}$ the eigenvalues of sample covariance matrix $\widehat{S}_n$, $n - 1 \geq p$. The fluctuation of the smallest eigenvalue $\lambda_{n,p}$ is characterized as follows:*

$$\frac{\log \lambda_{n,p} - \nu_{np}^-}{\delta_{np}^-} \xrightarrow{d} \mathcal{G}_1,$$

$$\nu_{np}^- = \log(\mu_{np}^-) + \frac{1}{8}(\delta_{np}^-)^2,$$

$$\delta_{np}^- = \frac{\sigma_{np}^-}{\mu_{np}^-},$$

where ν_{np}^- and δ_{np}^- are the centering and scaling parameters for $\log \lambda_{n,p}$ and $\mathcal{G}_1(x) = 1 - \mathcal{F}_1(-x)$ is the reflected Tracy–Widom law.

$\mu_{np}^- = \left(\sqrt{n_-} - \sqrt{p_-}\right)^2$, $\sigma_{np}^- = \left(\sqrt{n_-} - \sqrt{p_-}\right)\left(\frac{1}{\sqrt{p_-}} - \frac{1}{\sqrt{n_-}}\right)^{1/3}$, where $n_- = n - \frac{1}{2}$ and $p_- = p - \frac{1}{2}$, are the corrected centering and scaling constants for $\lambda_{n,p}$.

According to Ma (2012), the log-transform in the above Proposition 4.2.3 is justified by the superior numerical performance in the simulation study. Furthermore, the rationale behind this transformation is that for $\lambda_{n,p}$ the density function would

[3] The same theoretical result can be applied as well in the situation when $n < p$, just the roles of n and p should be reversed.

[4] $(n \wedge p) = \min(n, p)$.

be truncated at zero, and hence the asymptotic approximation by $\mathcal{G}_1$ would not be possible since the density of $\mathcal{G}_1$ is defined on the whole real line. In contrast, the distribution of the $\log \lambda_{n,p}$ is defined on the whole real line.

4.2.3 Marčenko and Pastur (1967) Result

The Marčenko–Pastur law describes the joint behavior of the p eigenvalues of a sample covariance matrix $\widehat{S}_n$. To be more precise, it characterizes the limiting distribution of the bulk spectrum and provides an accurate model for the sample eigenvalues in finite samples. This result establishes that the "joint limit" of the p sample eigenvalues converges to the Marčenko–Pastur distribution for a large class of population spectrum models.

A motivating example is presented in Fig. 4.2. This is a normalized histogram of the sample eigenvalues, where the data is drawn from the standard normal, and the number of observations is $n = 600$, while the dimensionality is $p = 150$.

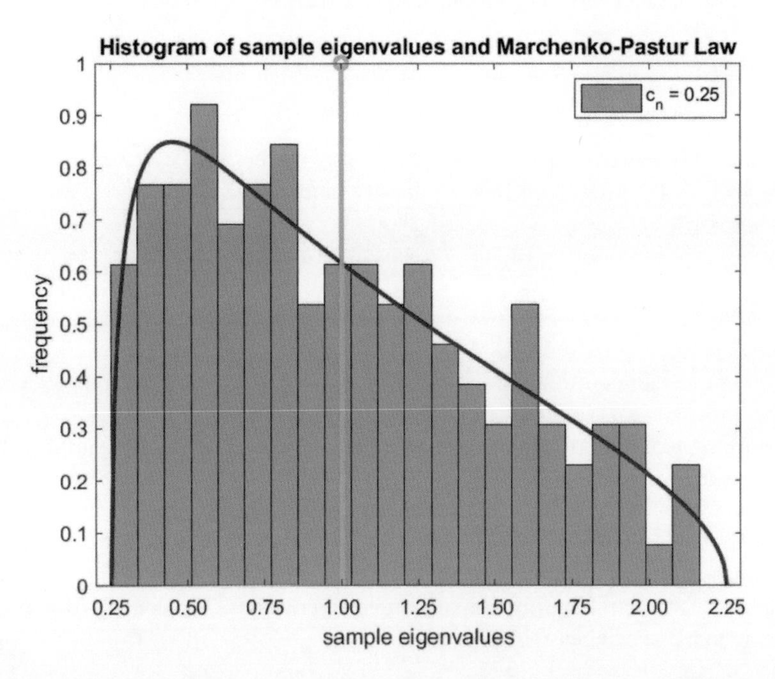

Fig. 4.2 The "overspreading" of sample eigenvalues. The normalized histogram of sample eigenvalues is generated using one realization of the data for $p = 150$, $n = 600$. The data is i.i.d Gaussian with mean zero and true population covariance matrix $\Sigma = I$. The green line is the limiting distribution function of population eigenvalues, which is a point mass at 1, i.e., δ_1. The red line is the limiting pdf $f_{c,\sigma^2}(x)$ of sample eigenvalues (Marčenko–Pastur density). Based on Figure 28.3 of Akemann et al. (2011)

According to traditional asymptotics, sample eigenvalues should converge to the unique population value, which is equal to one. However, as graph illustrates, this convergence result does not hold in the setup where p is comparable in magnitude to n. Furthermore, the red line that fits the shape of the histogram very well is the Marčenko–Pastur density function, which is defined later.

The limiting behavior of the p-dimensional vector of sample eigenvalues is mathematically challenging to estimate, as the size of the vector grows to infinity when $n \to \infty$ under high-dimensional asymptotics. A more convenient way to solve this problem is to consider the limit of the probability measure that is associated with the p-dimensional vector instead of addressing the vector itself. Thus, the following measure is considered as a convenient tool for the analysis of the "joint" behavior of the sample eigenvalues:

$$F_n(x) = \frac{1}{p} \sum_{i=1}^{p} \mathbb{1}\{\lambda_{n,i} \leq x\}, \ x \in \mathbb{R}^+,$$

which is the empirical distribution function of the sample eigenvalues.

Proposition 4.2.4 *Consider that the sample covariance matrix* $\widehat{S}_n = \frac{1}{n} X_n' X_n$. X_n *is an* $n \times p$ *matrix whose entries are i.i.d. with* $E(X_{ij}) = 0$, $V(X_{ij}) = \sigma^2$ *and* $p, n \to \infty$ *with* $c_n = p/n \to c$, $c \in [0, \infty)$. *Then almost surely the sequence* $F_n(x)$ *weakly converges to the Marčenko–Pastur law* F_{c,σ^2}.[5]

The Marčenko–Pastur distribution F_{c,σ^2} in the special case $\Sigma = \sigma^2 I$ as given above in Proposition 4.2.4 has the following density function:

$$f_{c,\sigma^2}(x) = \frac{1}{2\pi c \sigma^2} \frac{\sqrt{b_c - x}\sqrt{x - a_c}}{x} \mathbb{1}\{a_c \leq x \leq b_c\},$$

where $b_c = \sigma^2(1 + \sqrt{c})^2$ and $a_c = \sigma^2(1 - \sqrt{c})^2$.

$f_{c,\sigma^2}(x)$ has an additional point mass of value $1 - 1/c$ at the origin $x = 0$, if $c > 1$.

Returning to the motivating example in Fig. 4.2, we can observe that the sample eigenvalues are spread on the interval $[a_c, b_c] = [0.25, 2.25]$ for $c_n = 0.25$ which is the support of the Marčenko–Pastur density function.

The above result can be generalized to the arbitrary population covariance matrix Σ_n. However, in that case there is no closed-form solution for the density $f_{c,\sigma^2}(x)$ of the limiting spectral distribution of sample eigenvalues. The density form is only available in the special case of $\Sigma = \sigma^2 I$.

[5] Weak convergence in a sense $F_n(g) \to F(g)$, for any continuous and bounded function g, where F is the distribution function.

To consider a general result, we must first introduce the empirical distribution function of the population eigenvalues:

$$H_n(x) = \frac{1}{p} \sum_{i=1}^{p} \mathbb{1}\{\tau_{n,i} \leq x\}, \ x \in \mathbb{R}^+.$$

Proposition 4.2.5 *Consider the sample covariance matrix* $\widehat{S}_n = \frac{1}{n} \Sigma_n^{1/2} X_n' X_n \Sigma_n^{1/2}$. X_n *is an* $n \times p$ *matrix whose entries are i.i.d. with* $E(X_{ij}) = 0$, $V(X_{ij}) = 1$ *and* $p, n \to \infty$ *with* $c_n = p/n \to c$, $c \in (0, \infty)$.

Let Σ_n *be a sequence of non-negative definite Hermitian matrices of size* $p \times p$. *The sequence of* Σ_n *is either deterministic or independent of* $\widehat{S}_n$. *Furthermore, the sequence of the empirical spectral distributions* $\{H_n\}$ *of* Σ_n *weakly converges to a non-random limiting spectral distribution* H.

Then, almost surely, the sequence of the empirical spectral distributions $\{F_n\}$ *weakly converges to a non-random limiting distribution* $F_{c,H}$.

Moreover, the Stieltjes transform m *of* $F_{c,H}$ *is implicitly defined by the following fixed-point equation:*

$$m(z) = \int \frac{1}{\tau(1 - c - c\, z\, m(z)) - z} \, dH(\tau), \ z \in \mathbb{C}^+,$$

where $\mathbb{C}^+$ *denotes the set of complex numbers with strictly positive imaginary parts.*

The fixed-point equation above is proven to have a unique solution as a function from $\mathbb{C}^+$ to $\mathbb{C}^+$. Furthermore, $F_{c,H}$ is a generalized Marčenko–Pastur distribution, where the subscript (c, H) indicates the dependence on the limiting ratio c and on the limiting population spectral distribution H.

It is important to note that the almost sure weak convergence of the sequence F_n is implied by the almost sure convergence of the corresponding sequence of Stieltjes transforms $m_n(z) \to m(z)$ (see, e.g., Theorem 2.7, Yao et al. 2015). $m_n(z)$ is defined for the sample covariance matrix $\widehat{S}_n$ as $m_n(z) = \text{trace}[(\widehat{S}_n - zI)^{-1}]/p$, $z \in \mathbb{C}^+$.

Thus, the spectral distribution of $\widehat{S}_n$ is asymptotically non-random $F_{c,H}$ and has the Stieltjes transform m, which is a solution to the fixed-point equation.

Moreover, the limiting distribution of the sample eigenvalues $F_{c,H}$ can be recovered indirectly through its Stieltjes transform m as following:

$$m(z) = \int_{-\infty}^{+\infty} \frac{1}{\lambda - z} \, dF_{c,H}(\lambda), \ z \in \mathbb{C}^+.$$

Considering the two definitions for the Stieltjes transform m of $F_{c,H}$, we can observe that the Marčenko–Pastur equation in Proposition 4.2.5 links the limiting spectral distribution of the sample eigenvalues $F_{c,H}$ to the limiting spectral distribution of the population eigenvalues H through a given limiting ratio c.

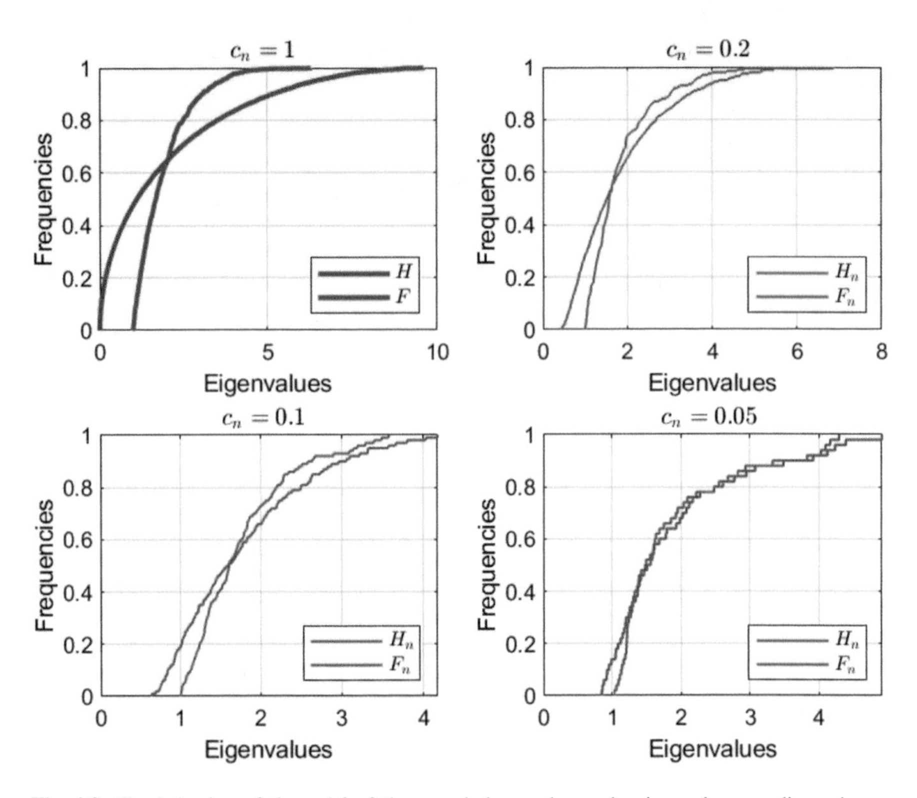

Fig. 4.3 The behavior of the e.d.f of the population and sample eigenvalues as dimension p increases. The population covariance matrix is $\Sigma_n = \text{diag}(\tau_{n,1}, \ldots, \tau_{n,p})$. The i-th population eigenvalue is equal to $\tau_{n,i} = H_n^{-1}((i - 0.5)/p)$, $i = 1, \ldots, p$, where the limiting spectral distribution H is given by the distribution of $1 + 10Z$, and $Z \sim Beta(1, 10)$. The subscript n in Σ_n emphasizes the fact that the population covariance matrix Σ_n depends on n through the dimensionality $p = p(n)$

Furthermore, the distribution of sample eigenvalues is considered to be a highly non-linear deformation of the population eigenvalues' distribution, and the Marčenko–Pastur equation formulates this deformation mathematically (Akemann et al. 2011). The latter statement is illustrated in Fig. 4.3.

The Marčenko–Pastur law holds under weak assumptions and in very wide generality, including the framework with spiked (generalized spiked) population covariance models, which were introduced in Sect. 2.4.

As the number of spikes k is assumed to be fixed in Johnstone's model under high-dimensional asymptotics, when $p, n \to \infty$ with $c_n = p/n \to c > 0$, the empirical spectral distribution F_n of the sample covariance matrix $\widehat{S}_n$ still converges to the Marčenko–Pastur law. However, the asymptotic behavior of the extreme (spiked) eigenvalues and the corresponding eigenvectors is modified. The latter is discussed in the subsequent sections.

4.2.4 *Spiked Population Model by Johnstone (2001)*

Motivated by the fact that many datasets in finance and statistical learning applications exhibit a few large eigenvalues that are separated from the bulk, Johnstone (2001) introduced the spiked covariance model, which can be considered as a finite-rank perturbation of the identity matrix. In other words, the base population covariance model is identity, but there are a few large "spiked" eigenvalues.

The general model considered under this framework is as follows: X_n is an $n \times p$ matrix whose entries are i.i.d. complex or real valued, with $E(X_{ij}) = 0$, $E(|X_{ij}|^2) = 1$, $E(|X_{ij}|^4) < \infty$, $p, n \to \infty$ with $c_n = p/n \to c$, $c \in (0, \infty)$.

The sample covariance matrix is defined as $\widehat{S}_n = \frac{1}{n}\Sigma^{1/2}X_n' X_n \Sigma^{1/2}$, where the spiked population covariance model is given in the form

$$\Sigma = \text{diag}(\tau_1, \ldots, \tau_k, 1, \ldots, 1),$$

with $\tau_1 > \tau_2 > \ldots > \tau_k > 1$. The population eigenvalues are not restricted to be unique and can have the respective multiplicities $m_1, \ldots, m_k$. In what follows, we denote as J_i the set of m_i indexes that correspond to τ_i.

The following almost sure bound results according to Baik and Silverstein (2006) hold for a general class of samples, with real and complex data, which is not necessarily Gaussian.

Proposition 4.2.6 *Let $p \to \infty$, $n = n(p) \to \infty$, $c_n = p/n \to c \in (0, 1)$. Let us consider the large fundamental spikes such that $\tau_i > 1 + \sqrt{c}$, and the small fundamental spikes such that $\tau_i < 1 - \sqrt{c}$. Then, for the corresponding sample eigenvalues $\lambda_{n,i}$ the following holds:*

(i) Large fundamental spikes $\tau_i > 1 + \sqrt{c}$:

$$\lambda_{n,i} \to \tau_i \left(1 + \frac{c}{\tau_i - 1}\right) \; a.s. \text{ for } i \in J_i,$$

(ii) Large non-fundamental spikes $1 < \tau_i \leq 1 + \sqrt{c}$:

$$\lambda_{n,i} \to (1 + \sqrt{c})^2 \; a.s. \text{ for } i \in J_i,$$

(iii) Small non-fundamental spikes $1 - \sqrt{c} \leq \tau_i < 1$:

$$\lambda_{n,i} \to (1 - \sqrt{c})^2 \; a.s. \text{ for } i \in J_i,$$

(iv) Small fundamental spikes $\tau_i < 1 - \sqrt{c}$:

$$\lambda_{n,i} \to \tau_i \left(1 + \frac{c}{\tau_i - 1}\right) \; a.s. \text{ for } i \in J_i.$$

The special cases for $c = 1$ and $c > 1$ are detailed in Corollaries C.1.1–C.1.2 (Appendix C.1) for the sake of saving space. The results are similar, but with minor adjustments.

Proposition 4.2.6 demonstrates that no sample eigenvalue $\lambda_{n,i}$ has an almost sure limit outside of the support of the Marčenko–Pastur density $[(1-\sqrt{c})^2, (1+\sqrt{c})^2]$ if the population eigenvalue τ_i is in the interval $[(1-\sqrt{c}), (1+\sqrt{c})]$. This interval characterizes the quantitative measure of population eigenvalue τ_i being sufficiently close to one, such that the Marčenko–Pastur density is not affected by the spike. The large non-fundamental spikes, $1 < \tau_i \leq 1+\sqrt{c}$, and the small ones, $1-\sqrt{c} \leq \tau_i < 1$, generate the sample eigenvalues that are not separated in the limit from the bulk that corresponds to the unit population eigenvalues $\tau_i = 1$, and thus converge almost surely to the right $(1+\sqrt{c})^2$ and left $(1-\sqrt{c})^2$ edges of the Marčenko–Pastur distribution, respectively. This is a well-known property: The sample eigenvalues associated with the distinct population eigenvalue clusters are merged together in the limit, making them indistinguishable (El Karoui 2008). Such a phenomenon is also observable in the null case when there are no spikes.

An illustration of this is provided in Fig. 4.4. In this example, the distribution of population eigenvalues H is uniform on the set $\{1, 4, 10\}$, i.e.,

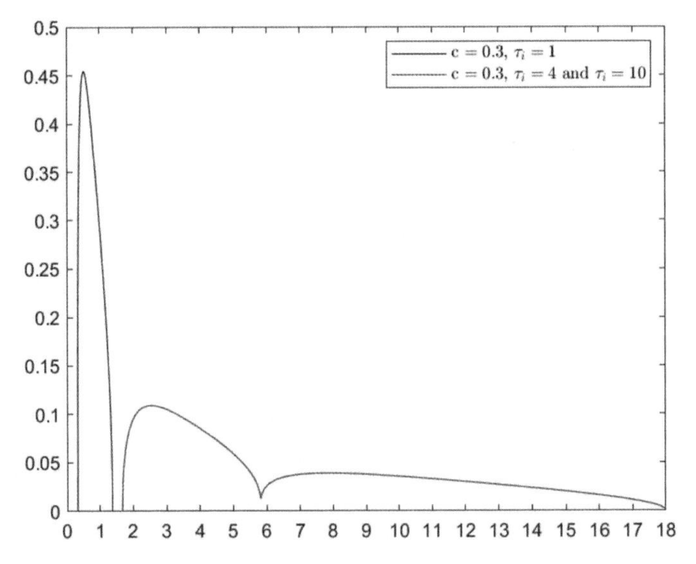

Fig. 4.4 The density function for the limiting spectral distribution $F_{c,H}$ of sample eigenvalues, where $c = 0.3$ and H is uniform on the set $\{1, 4, 10\}$. The sample eigenvalues that correspond to the clusters of the population eigenvalues $\tau_i = 4$ and $\tau_i = 10$ are merged. This is indicated by the red line, while the blue line demonstrates that the sample eigenvalues that correspond to the unit population eigenvalues $\tau_i = 1$ are well-separated (The density function is derived using the Matlab codes denfun03.m, solfun02.m, netfun02.m and supfun07.m of the QuEST package by Ledoit and Wolf (2017)). Based on Figure 2.4 of Yao et al. (2015)

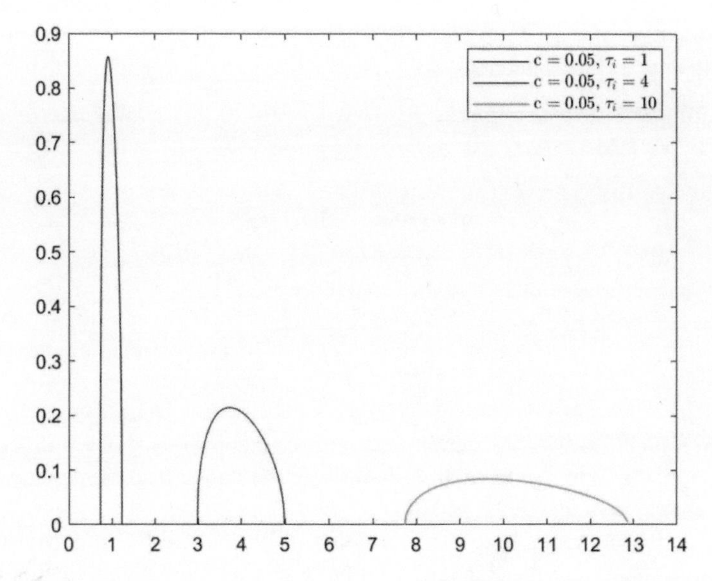

Fig. 4.5 The density function for the limiting spectral distribution $F_{c,H}$ of sample eigenvalues, where $c = 0.05$ and H is uniform on the set $\{1, 4, 10\}$. The sample eigenvalues that correspond to the distinct population eigenvalue clusters $\tau_i = 1$, $\tau_i = 4$, and $\tau_i = 10$ are well-separated. This is indicated by different lines and colors

$H = \{\underbrace{10, \ldots, 10}_{p/3}, \underbrace{4, \ldots, 4}_{p/3}, \underbrace{1, \ldots, 1}_{p/3}\}$, and the sample eigenvalues that correspond to the population values 4 and 10 are blurred together in a single large cluster in the limit when $c = 0.3$. However, as the dimension-to-sample size ratio c gets closer to zero, and thus the limiting behavior of the sample eigenvalues is close to the traditional asymptotics, the distinct clusters are well-separated in the limit. This is demonstrated in Fig. 4.5.

In contrast, if the population eigenvalue τ_i is outside of the interval $[(1 - \sqrt{c}), (1 + \sqrt{c})]$, meaning that in the limit it is no longer close to one, the corresponding sample eigenvalue is outside of the support of the Marčenko–Pastur density, $[(1 - \sqrt{c})^2, (1 + \sqrt{c})^2]$.

Furthermore, each population eigenvalue τ_i outside of the interval $[(1 - \sqrt{c}), (1 + \sqrt{c})]$ is considered to be a spike and changes the limiting value of the sample eigenvalue $\lambda_{n,i}$ to be equal to $\tau_i \left(1 + \frac{c}{\tau_i - 1}\right)$, thus, pulling it from the support of the Marčenko–Pastur density.

This is the so-called phase transition phenomenon that was first introduced by Baik et al. (2005) to describe the behavior of the largest sample eigenvalue. This phenomenon basically means that no large spike can be detected below the threshold $1 + \sqrt{c}$ and no small spike can be detected above the threshold $1 - \sqrt{c}$.

The threshold $1 + \sqrt{c}$ also plays a key role in the distributional result derived by Paul (2007) for the sample eigenvalues that correspond to the large fundamental spikes in the case of $c \in (0, 1)$. The following statement holds under an assumption of the real valued Gaussian data.

Proposition 4.2.7 *Suppose that* $\tau_i > 1 + \sqrt{c}$, $c \in (0, 1)$ *and has multiplicity 1. Then, as* $p, n \to \infty$ *so that* $c_n = p/n - c = o(n^{-1/2})$,

$$\sqrt{n}(\lambda_{n,i} - \rho(\tau_i)) \xrightarrow{d} \mathcal{N}(0, \sigma^2(\tau_i)),$$

$$\rho(\tau) = \tau\left(1 + \frac{c}{\tau - 1}\right),$$

$$\sigma^2(\tau) = 2\tau^2\left(1 - \frac{c}{(\tau - 1)^2}\right).$$

Hence, if the dimension-to-sample size ratio p/n converges to the limiting ratio c fast enough, the Gaussian fluctuation result with the standard convergence rate $\sqrt{n}$ holds for the large fundamental spikes.

This result is comparable to the traditional result derived by Anderson (1963) (see, Sect. 2.2.3). Under the standard asymptotics, when the i-th population eigenvalue τ_i has multiplicity 1, the i-th sample eigenvalue $\lambda_{n,i}$ is asymptotically $\mathcal{N}(\tau_i, (1/n)2\tau_i^2)$.

In the high-dimensional case, however, the sample eigenvalues are estimated with a bias, although the variance shrinks. This is accounted for through the concentration ratio c that appears in the centering and scaling constants, $\rho(\tau)$ and $\sigma(\tau)$, respectively.

Table 4.1 presents the results of the Monte Carlo study for sample eigenvalues in the spiked population model setup. The outputs of the Monte Carlo study confirm the theoretical findings of Proposition 4.2.6. Further results for $c_n = 0.2$ and $c_n = 0.5$ are provided in Tables B.8 and B.9 (Appendix B).

Here, as predicted by the theoretical findings, the first spiked eigenvalue τ_1 is estimated by the sample eigenvalue $\lambda_{n,1}$ with a bias. The second eigenvalue τ_2 is the non-fundamental spike and is exactly equal to the threshold $1 + \sqrt{c_n}$, and thus the corresponding sample eigenvalue $\lambda_{n,2}$ tends to $(1 + \sqrt{c_n})^2$. The sample eigenvalue $\lambda_{n,3}$ that corresponds to the first unit population eigenvalue in the spectrum, i.e., $\tau_3 = 1$, also converges to the value $(1 + \sqrt{c_n})^2$. Hence, as demonstrated in Fig. 4.4, for the large values of c, the sample eigenvalues that correspond to the non-fundamental spikes are not distinguishable from the right edge of the bulk. The sample eigenvalue $\lambda_{n,p}$ that corresponds to the last unit population eigenvalue in the spectrum, i.e., $\tau_p = 1$, converges to the left edge of the bulk, i.e., $(1 - \sqrt{c_n})^2$.

Figures 4.6 and 4.7 illustrate the distributional result derived by Paul (2007) for cases when the largest population eigenvalue τ_1 is equal to 2 and 10. Figures A.5, 4.6, and A.7 in Appendix A present the same graphs for the cases when τ_1 is equal to 3, 4, and 5.

An extension to the case when the base population covariance matrix is $\sigma^2 I$ means that the spiked population covariance matrix is in turn $\Sigma = \text{diag}(\tau_1, \ldots, \tau_k, \sigma^2, \ldots, \sigma^2)$, with $\tau_1 > \tau_2 > \ldots > \tau_k > \sigma^2$. The phase

Table 4.1 Properies of sample eigenvalues. Spiked covariance matrix

$c_n = 0.8\ (p = 200, n = 250)$				
$\tau_1 = 2$	$\tau_1 = 3$	$\tau_1 = 4$	$\tau_1 = 5$	$\tau_1 = 10$
		Mean($\lambda_{n,1}$)		
3.66	4.21	5.07	6.01	10.85
(0.1134)	(0.2307)	(0.3391)	(0.4476)	(0.9152)
		Theoretical S.E.s		
(0.0800)	(0.2400)	(0.3415)	(0.4359)	(0.8900)
		$\widehat{\text{Bias}}(\lambda_{n,1})$		
1.66	1.21	1.07	1.01	0.85
		Theoretical Bias($\lambda_{n,1}$)		
1.60	1.20	1.07	1.00	0.89
		$\tau_2 = 1 + \sqrt{c_n}$		
		Mean($\lambda_{n,2}$)		
3.48	3.54	3.55	3.55	3.56
(0.0772)	(0.0907)	(0.0924)	(0.0907)	(0.0954)
		$\tau_3 = 1$		
		Mean($\lambda_{n,3}$)		
3.37	3.40	3.40	3.40	3.40
(0.0621)	(0.0634)	(0.0662)	(0.0659)	(0.0662)
		$\tau_p = 1$		
		Mean($\lambda_{n,p}$)		
0.01	0.01	0.01	0.01	0.01
(0.0018)	(0.0018)	(0.0018)	(0.0019)	(0.0019)

Note: $p = 200, n = 250, c_n = 0.8$. The population covariance matrix is $\Sigma = \text{diag}(\tau_1, \tau_2, 1 \ldots, 1)$ τ_1 takes values in the set $\{2, 3, 4, 5, 10\}$, $\tau_2 = 1 + \sqrt{c_n}$, and $\tau_3 = \ldots = \tau_p = 1$. The number of Monte Carlo replications is $R = 1000$

Remark: for $c_n = 0.8$ the limiting values are $(1 + \sqrt{c_n})^2 = 3.5889$ and $(1 - \sqrt{c_n})^2 = 0.0111$, respectively. The theoretical bias and standard deviation are computed according to Proposition 4.2.7

transition phenomenon also takes place in this setup, however, if the population eigenvalues are outside of the adjusted interval $[\sigma^2(1 - \sqrt{c}), \sigma^2(1 + \sqrt{c})]$. The same bound and distributional results hold as before, where the only change takes place in the "transition" threshold value that is corrected to be $\sigma^2(1 + \sqrt{c})$ for large spikes and $\sigma^2(1 - \sqrt{c})$ for small ones. These generalizations are useful in many applications, one of them considered later is the partial sphericity test.

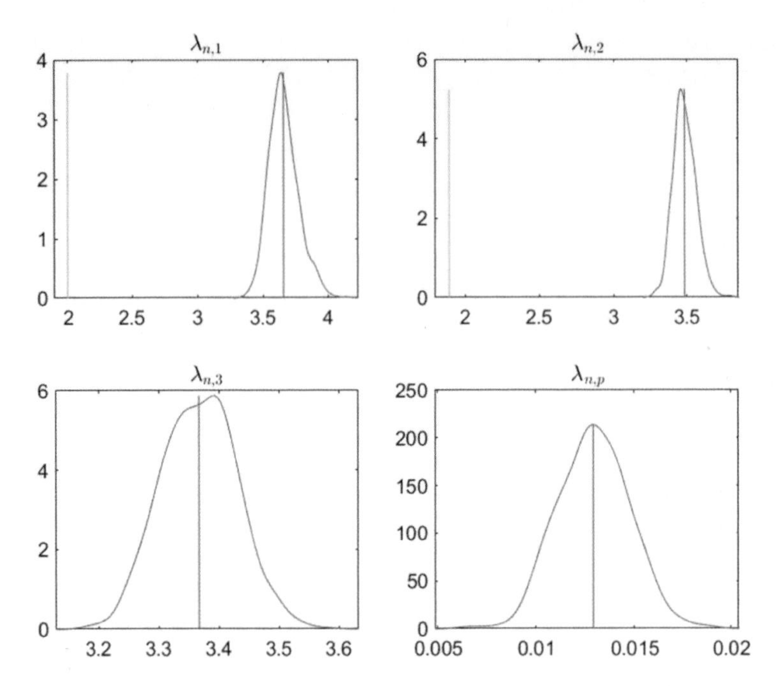

Fig. 4.6 $p = 200$, $n = 250$, $c_n = 0.8$. The population covariance matrix is $\Sigma = \text{diag}(\tau_1, \tau_2, 1, \ldots, 1)$. $\tau_1 = 2$, $\tau_2 = 1 + \sqrt{c_n}$ and $\tau_3 = \ldots = \tau_p = 1$. The number of Monte Carlo replications is $R = 1000$. The densities for $\lambda_{n,1}$, $\lambda_{n,2}$, $\lambda_{n,3}$, and $\lambda_{n,p}$ are estimated with the "Normal" kernel density estimator

4.2.5 Generalized Spiked Population Model of Bai and Yao (2012)

The generalized spiked population model is an extended version of Johnstone's spiked population model, with the main difference being in the base covariance model, which is no longer anymore to be an identity, but rather is of arbitrary form. The model can be considered to be a finite-rank perturbation of a general covariance matrix that has the population eigenvalues $\beta_{n,j}$s.

Hence, the spectrum of the spiked covariance model consists of two clusters:

$$\text{spec}(\Sigma_n) = \{\tau_1, \ldots, \tau_k, \beta_{n,1}, \ldots, \beta_{n,p-K}\},$$

where the number of spike population eigenvalues $\tau_1, \ldots, \tau_k$ is fixed and the spikes are finite. Moreover, the spike population eigenvalues $\tau_1, \ldots, \tau_k$ are not necessarily larger than the base population eigenvalues $\beta_{n,1}, \ldots, \beta_{n,p-K}$ as in Johnstone's model and can have respective multiplicities $m_1, \ldots, m_k$ larger than one, where $m_1 + \ldots + m_k = K$.

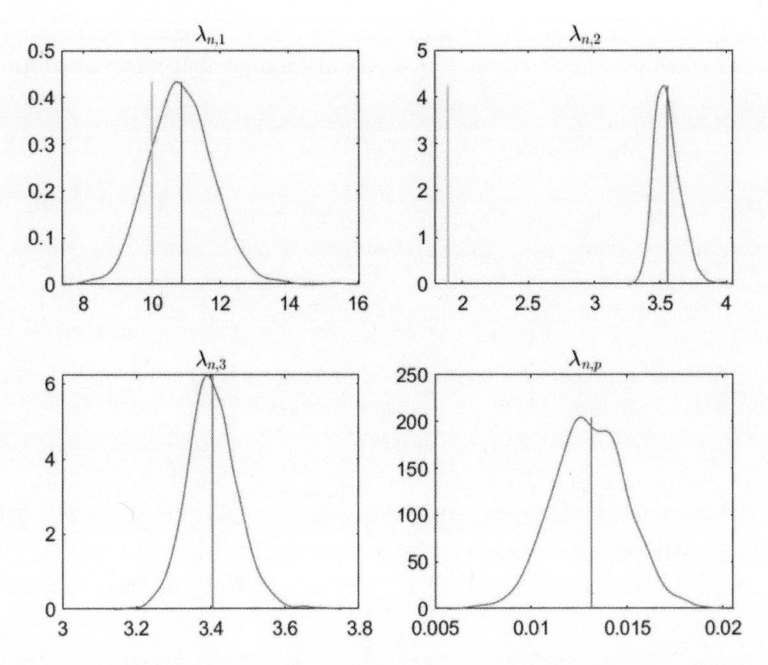

Fig. 4.7 $p = 200$, $n = 250$, and $c_n = 0.8$. The population covariance matrix is $\Sigma = \text{diag}(\tau_1, \tau_2, 1 \ldots, 1)$. $\tau_1 = 10$, $\tau_2 = 1 + \sqrt{c_n}$, and $\tau_3 = \ldots = \tau_p = 1$. The number of Monte Carlo replications is $R = 1000$. The densities for $\lambda_{n,1}$, $\lambda_{n,2}$, $\lambda_{n,3}$, and $\lambda_{n,p}$ are estimated with the "Normal" kernel density estimator

The limiting spectral distribution H of the base population eigenvalues $\beta_{n,1}, \ldots, \beta_{n,p-K}$ is arbitrary, in contrast, to the standard case, where it is assumed to be a point mass at one, i.e., δ_1.

Furthermore, the limiting spectral distribution $F_{c,H}$ of the sample covariance matrix $\widehat{S}_n$ is not affected by the perturbation in the population model, since the number of spikes is fixed and does not depend on $p = p(n)$, so that in the limit, the point mass of the sample eigenvalues that corresponds to the spikes is vanishing.

Thus, the limiting spectral distribution of the sample covariance matrix $\widehat{S}_n$ still converges to the generalized Marčenko–Pastur law $F_{c,H}$.

The generalized spikes $\tau_1, \ldots, \tau_k$ are assumed to be well-separated from the base population eigenvalues $\beta_{n,1}, \ldots, \beta_{n,p-K}$. Moreover, they pull the corresponding sample eigenvalues outside the support of the generalized Marčenko–Pastur distribution $F_{c,H}$. The sample eigenvalues that correspond to the spikes are highly non-linear transformations of the population counterparts.

The transformation function ψ that describes the link between the population spikes $\tau_1, \ldots, \tau_k$ and the respective set of the sample eigenvalues is closely related to the Stieltjes transform m of the limiting spectral distribution $F_{c,H}$.

It should be noted that the Stieltjes transform m fully characterizes the limiting spectral distribution $F_{c,H}$, and hence its support, according to the inversion formula:

$$f_{c,H}(x) = \frac{1}{\pi} \lim_{\varepsilon \to 0} \operatorname{Im}[m(x + i\varepsilon)],$$

where Im[·] denotes the imaginary part of a complex number.

Furthermore, Silverstein and Choi (1995) formulated how the support of the generalized Marčenko–Pastur distribution $F_{c,H}$ relates to the support of the limiting spectral distribution of the population eigenvalues H in terms of its Stieltjes transform m.

The corresponding theoretical result, however, is given in terms of $\underline{F}_{c,H}$ and $\underline{m}$ for the sake of convenience. The generalized Marčenko–Pastur distribution $\underline{F}_{c,H}$ is defined for the *companion* sample covariance matrix $\underline{\widehat{S}_n}$. Further details with regard to the *companion* sample covariance matrix $\underline{\widehat{S}_n}$, its limiting spectral distribution $\underline{F}_{c,H}$, and the respective Stieltjes transform $\underline{m}$, together with the corresponding links to $\widehat{S}_n$, $F_{c,H}$, and m are provided in Appendix C.2.

In the following equations, we denote the support of H as Γ_H and the support of $\underline{F}_{c,H}$ as $\Gamma_{\underline{F}_{c,H}}$ ($\Gamma_{F_{c,H}}$ is the support of $F_{c,H}$).

Proposition 4.2.8 *If* $\lambda \notin \Gamma_{\underline{F}_{c,H}}$, *then* $\underline{m}(\lambda) \neq 0$ *and* $\tau = -\dfrac{1}{\underline{m}(\lambda)}$ *satisfies:*

(i) $\tau \notin \Gamma_H$ *and* $\tau \neq 0$;
(ii) $\psi'(\tau) > 0$.

Conversely, if τ *satisfies (i)–(ii), then* $\lambda = \psi(\tau) \notin \Gamma_{\underline{F}_{c,H}}$.

Hence, according to the above result, if the non-zero population eigenvalue τ is outside the support of the limiting spectral distribution H (i.e., τ is spiked) and satisfies the condition $\psi'(\tau) > 0$, then the corresponding sample eigenvalue λ is a non-linear transformation of the population eigenvalue τ, $\lambda = \psi(\tau)$, and is outside the support of $\Gamma_{\underline{F}_{c,H}}$ and, respectively, $\Gamma_{F_{c,H}}$.

Thus, the spike population eigenvalue τ is proven to pull the corresponding sample counterpart λ outside the support of $F_{c,H}$, and the exact form of the non-linear transformation is provided, where function $\psi(\cdot)$ is of the following form:

$$\psi(\tau) = \psi_{c,H}(\tau) = \tau + c \int \frac{t\tau}{\tau - t} \, dH(t).$$

Furthermore, Proposition 4.2.8 provides a tool to determine the spike population eigenvalues by checking the condition $\psi'(\tau) > 0$, where

$$\psi'(\tau) = 1 - c \int \frac{t^2}{(\tau - t)^2} \, dH(t).$$

The below proposition (from Bai & Yao 2012) states the bound results for sample eigenvalues that correspond to the spikes in the generalized model based on the condition $\psi'(\tau) > 0$ and the definition of the function ψ.

Proposition 4.2.9 *Let $p \to \infty$, $n = n(p) \to \infty$, $c_n = p/n \to c > 0$. Under the generalized spiked population model,*

(i) For a fundamental spike eigenvalue τ_i satisfying $\psi'(\tau) > 0$,

$$\lambda_{n,i} \to \psi_i = \psi(\tau_i) \; a.s. \text{ for } i \in J_i,$$

(ii) For a non-fundamental spike eigenvalue τ_i satisfying $\psi'(\tau_i) \leq 0$,

$$\lambda_{n,i} \to \alpha_i, \; a.s. \text{ for } i \in J_i,$$

where α_i is the α-th quantile of $F_{c,H}$, where $\alpha = H(0, \tau_i]$.

Proposition 4.2.9 separates spike eigenvalues τ into two groups: those with a positive derivative, $\psi'(\tau) > 0$, are the fundamental spikes, and those with a non-positive derivative, $\psi'(\tau) \leq 0$, are the non-fundamental spikes.

The fundamental spike τ_i is defined to be such that for large enough n, exactly m_i sample eigenvalues $\lambda_{n,i}$ will cluster in a vicinity of $\psi_{c,H}(\tau_i)$, which is outside the support of the limiting spectral distribution $F_{c,H}$.

The non-fundamental spike τ_i is, on the contrary, such that the sample counterpart $\lambda_{n,i}$ is not distinguishable from the bulk in the limit and, hence, converges almost surely to the α quantile of the limiting spectral distribution $F_{c,H}$, where the quantile depends on the population eigenvalue τ_i through the limiting spectral distribution H as follows: $\alpha = H(0, \tau_i]$.

A graphical illustration of the bound results in the generalized spiked population model is provided in Fig. 4.8. The distribution of the base population eigenvalues $\beta_{n,j}$'s, or in other words H, is uniform on the set $\{1, 4, 10\}$ such that

$$H = \{\underbrace{10, \ldots, 10}_{200}, \underbrace{4, \ldots, 4}_{200}, \underbrace{1, \ldots, 1}_{200}\}.$$

Furthermore, there are four spike population eigenvalues $\tau_1 = 15$, $\tau_2 = 6$, $\tau_3 = 2$, and $\tau_4 = 0.05$, each of multiplicity one. Thus, the spectrum of the spiked covariance model Σ_n is given as follows:

$$\text{spec}(\Sigma_n) = \{15, \underbrace{10, \ldots, 10}_{200}, 6, \underbrace{4, \ldots, 4}_{200}, 2, \underbrace{1, \ldots, 1}_{200}, 0.5\}.$$

By checking the condition $\psi'(\tau) > 0$ for each of the spikes, we can conclude that $\tau_1 = 15$, $\tau_3 = 2$, and $\tau_4 = 0.05$ are the fundamental spikes as the derivative is positive, while the spike $\tau_2 = 6$ is non-fundamental, as the derivative is negative. Thus, the sample counterparts that correspond to the three fundamental spikes lie outside the support of generalized Marčenko–Pastur distribution $F_{c,H}$, while the

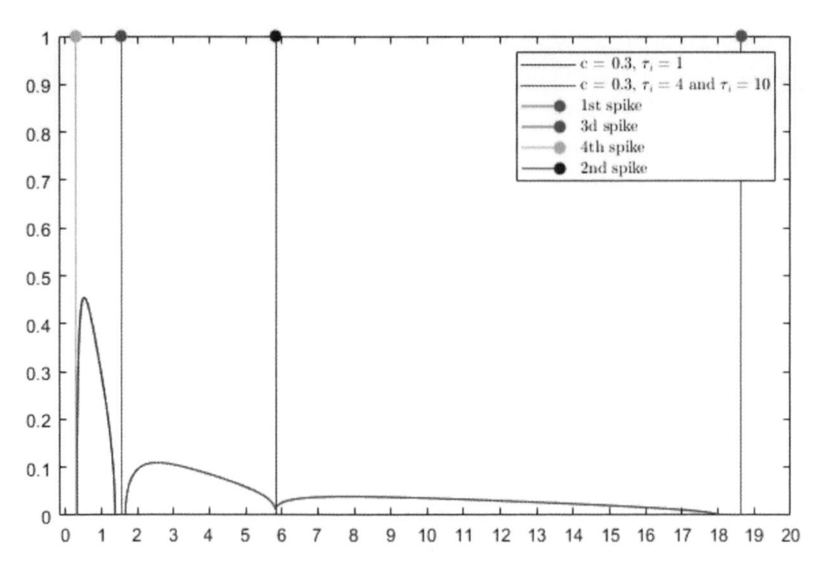

Fig. 4.8 The density function for the limiting spectral distribution $F_{c,H}$ of the sample eigenvalues, where $c = 0.3$. H is uniform on the set $\{1, 4, 10\}$. The blue line corresponds to the limiting value associated with the first spike $\tau_1 = 15$, the purple line to $\tau_2 = 6$, the red line to $\tau_3 = 2$, and the yellow line to $\tau_4 = 0.05$ (The scaled parts of the graph are provided additionally in Figs. A.8, A.9, and A.10 in Appendix A)

sample eigenvalue that corresponds to the second spike, $\tau_2 = 6$, converges almost surely to the $\alpha = 2/3$ quantile of $F_{c,H}$.

Moreover, Fig. 4.8 illustrates that the limiting spectral distribution of the sample eigenvalues $F_{c,H}$ is not affected by the presence of the spiked eigenvalues in the spectrum of the population covariance matrix. The shape of the density function for the limiting ratio $c = 0.3$ resembles the one in Fig. 4.4.

It should be noted that from the definition of $\psi(\cdot)$, it is evident that when c is close to zero, and thus the traditional asymptotics framework applies, $\psi_{c,H}(\tau_i) \simeq \tau_i$. This means that any spike τ_i behaves as the fundamental spike under the low-dimensional setup. Moreover, under this traditional asymptotics framework the usual consistency property is recovered; the sample eigenvalues converge to the corresponding population eigenvalues. Thus, it is possible to reproduce the traditional asymptotics results using function $\psi(\cdot)$ in the special case of $c \to 0$.

The fluctuation result for the sample eigenvalues that correspond to the fundamental spikes is also available for the generalized setup. As under Johnstone's spiked population model, the Gaussian asymptotic distributional result is possible only under the assumption of the real Gaussian data and only for the spikes of multiplicity one. In general, when the spikes are of a higher multiplicity, the joint distribution of the respective sample counterparts is not Gaussian. In what follows, we present the theoretical result for the aforementioned restricted case. The more general case is considered, for the sake of saving space, in Appendix C.3.

However, it is worth noting the link between the traditional asymptotics result and the high-dimensional one. Under the high-dimensional asymptotics, if the entries of the data are real Gaussian, the sample eigenvalues that correspond to the same spiked population counterpart converge in distribution to the eigenvalues of the Gaussian Wigner matrix. However, the same result was obtained under the traditional asymptotics in the case of sample eigenvalues that represent the unit population counterparts (see, e.g., Proposition 2.2.4). One can perceive the similarity in these results, as in both settings, the sample eigenvalues are associated with the same population values.

The following proposition is from Bai and Yao (2008).

Proposition 4.2.10 *Let $p \to \infty$, $n = n(p) \to \infty$, $c_n = p/n \to c > 0$. Under the generalized spiked population model, if $\psi'(\tau) > 0$ and τ_i is of multiplicity 1, then the limiting distribution is*

$$\sqrt{n}(\lambda_{n,i} - \psi(\tau_i)) \xrightarrow{d} \mathcal{N}(0, \sigma^2(\tau_i)),$$
$$\sigma^2(\tau_i) = 2\tau_i^2 \psi'(\tau_i),$$

where the functions $\psi(\tau)$ and $\psi'(\tau)$ were defined earlier.

4.3 Sample Eigenvectors

In the null case, when there are no spikes in the spectrum of the population covariance matrix, under the Gaussian assumption imposed on the entries of the data, the matrix of sample eigenvectors, $U_n = (u_{n,1}, \ldots, u_{n,p})$, is Haar distributed in the high-dimensional setup. This is the same result as under traditional asymptotics. However, in the latter setting, the sample eigenvectors correspond to the unit population eigenvalues, while in the high-dimensional framework, the population eigenvalues in the null case are not necessarily unit-valued. The result for the sample eigenvectors does not change when moving from one type of asymptotics to another as the joint distributions of the data entries are invariant under the orthogonal transformations in both asymptotic regimes. However, if the data entries are not Gaussian, the sample eigenvectors' matrix U_n is *asymptotically* Haar distributed in the high-dimensional framework, in other words, for large p, the distribution of eigenmatrix U_n is "close" to Haar distribution. This limiting type of behavior for eigenmatrix was first conjectured in Silverstein (1979). While there are many universality results for the sample eigenvalues under the high-dimensional asymptotics (e.g., Johnstone & Paul 2018), the corresponding results for the sample eigenmatrices are hard to establish because the dimensions of matrices are growing. In particular, it is hard to prove the above conjecture, as it is difficult to formulate the terminology "asymptotically Haar distributed" when the dimension p is increasing. The approach used to prove the asymptotic Haar distribution of the sample eigenvectors' matrix is based on the method created by Silverstein (1981). The idea of this approach is to show that if the sample eigenmatrix U_n is asymptotically Haar distributed, then the eigenvector $u_{n,i}$ (for

$i = 1, \ldots, p$) should be asymptotically uniform, and the respective stochastic process defined for $u_{n,i}$ as follows:

$$Y_p(t) = \sqrt{\frac{p}{2}} \sum_{i=1}^{[pt]} \left(|u_{n,i}|^2 - \frac{1}{p} \right)$$

should converge in distribution to a Brownian bridge when p converges to infinity.[6] Using this technique Bai et al. (2007) studied the properties of individual eigenvectors through the corresponding stochastic processes. They proved that if the first four moments of the data distribution match those of the standard Gaussian, then the sample eigenmatrix is asymptotically Haar distributed in case of a general (non-negative definite) population covariance matrix. Thus, they extended the original result for an identity covariance matrix introduced by Silverstein (1981). Recently, Bai et al. (2011) introduced the generalized approach to further prove the asymptotic Haar distribution for eigenmatrices.

In contrast, the limiting behavior of the sample eigenvectors in the non-null case is determined by the phase transition phenomenon. The following bound and distributional results for the sample eigenvectors are provided under the assumption of Johnstone's spiked population model (Paul 2007).

Proposition 4.3.1 *Let p, $n \to \infty$ such that $c_n = p/n \to c \in (0, 1)$. Let $\tilde{e}_i$ denote the $p \times 1$ vector with 1 in the i-th coordinate and zeros elsewhere, and let $u_{n,i}$ denote the eigenvector of sample covariance matrix $\widehat{S}_n$ associated with the eigenvalue $\lambda_{n,i}$.*

(i) If $\tau_i > 1 + \sqrt{c}$ and of multiplicity one,

$$|\langle u_{n,i}, \tilde{e}_i \rangle| \to \sqrt{\left(1 - \frac{c}{(\tau_i - 1)^2}\right) / \left(1 + \frac{c}{\tau_i - 1}\right)} \qquad a.s.$$

(ii) If $\tau_i \leq 1 + \sqrt{c}$,

$$\langle u_{n,i}, \tilde{e}_i \rangle \to 0 \qquad a.s.$$

Hence, Proposition 4.3.1 demonstrates that the phase transition phenomenon also takes place in the asymptotic behavior of the angle between the true $\tilde{e}_i$ and estimated eigenvectors $u_{n,i}$ associated with the non-unit population eigenvalue τ_i.

If the population eigenvalue τ_i has multiplicity one and $1 < \tau_i \leq 1 + \sqrt{c}$, then the cosine of the angle between the corresponding true and estimated eigenvectors converges almost surely to zero. However, if $\tau_i > 1 + \sqrt{c}$, the cosine of the angle converges almost surely to a positive limit that is not equal to one. This means that

[6] $[x]$ denotes the greatest integer smaller or equal to x.

asymptotically space spanned by the estimated eigenvectors $u_{n,i}$ does not coincide with space spanned by the true eigenvectors $\tilde{e}_i$. Thus, Proposition 4.3.1 proves once more the inconsistency of the result for the sample eigenvectors under the high-dimensional asymptotics, which was first derived by Johnstone and Lu (2004).

For the distributional result given below, the i-th sample eigenvector is expressed as $u_{n,i} = (u'_{n,iA}, u'_{n,iB})'$, where $u_{n,iA}$ is the sub-vector that corresponds to the first k coordinates. The vector e_i denotes the i-th canonical basis vector in $\mathbb{R}^k$.

Proposition 4.3.2 *Suppose that $\tau_i > 1 + \sqrt{c}$ and that τ_i has multiplicity 1. Then, as $p,\ n \to \infty$ such that $p/n - c = o(n^{-1/2})$,*

$$\sqrt{n}\left(\frac{u_{n,iA}}{||u_{n,iA}||} - e_i\right) \xrightarrow{d} \mathcal{N}_k(0, \Sigma(\tau_i)),$$

$$\Sigma(\tau_i) = \frac{\tau_i}{1 - \dfrac{c}{(\tau_i - 1)^2}} \cdot \sum_{1 \le l \ne i \le k} \frac{\tau_l}{(\tau_l - \tau_i)^2} e_l e_l',$$

where e_l denotes l-th canonical basis vector in $\mathbb{R}^k$.

The distributional result in Proposition 4.3.2 demonstrates that the sub-vector of the sample eigenvector $u_{n,iA}$ that corresponds to the first k sample eigenvalues (which in turn correspond to k population spikes) is asymptotically multivariate normal. This fluctuation result can be compared to the traditional result (Anderson 1963, see Sect. 2.3). However, unlike in the traditional asymptotics framework, in the high-dimensional case, the estimated sub-eigenvectors $u_{n,iA}$ do not converge to the population counterpart (namely, the respective sub-vector of $\tilde{e}_i$); they instead converge to the vector $e_i \in \mathbb{R}^k$, which is the eigenvector corresponding to $\rho(\tau_i) = \tau_i\left(1 + \dfrac{c}{\tau_i - 1}\right)$. This eigenvalue is the asymptotic limit of spiked sample eigenvalue $\lambda_{n,i}$ (see Proposition 4.2.7). Furthermore, the asymptotic covariance matrix has an additional term $\dfrac{(\tau_i - 1)^2}{(\tau_i - 1)^2 - c}$, depending on the concentration ratio c and the size of the spike τ_i.

Overall, the result illustrates that when τ_i is above the transition threshold $1 + \sqrt{c}$ and p/n converges to c fast enough, it is possible to give an approximation of the behavior of sample sub-eigenvectors associated with the "signal" part (spiked population eigenvalues).

The following finite sample result for the sub-vector corresponding to the last $p - k$ sample eigenvalues $u_{n,iB}$ (population eigenvalues are equal to one) resembles the conditional Haar distribution for the eigenvectors of the identity population covariance matrix (see, Sect. 2.3 and Propositions 2.3 and 2.3.4).

Proposition 4.3.3 *The vector $\dfrac{u_{n,iB}}{||u_{n,iB}||}$ is distributed uniformly on the unit sphere $\mathbb{S}^{p-k-1}$ and is independent of $||u_{n,iB}||$.*

Table 4.2 Properies of sample eigenvectors. Spiked covariance matrix.

$c_n = 0.8$ ($p = 200, n = 250$)				
$\tau_1 = 2$	$\tau_1 = 3$	$\tau_1 = 4$	$\tau_1 = 5$	$\tau_1 = 10$
		$\cos\phi(u_{n,1}, e_1)$		
0.32	0.74	0.84	0.89	0.95
(0.1660)	(0.0595)	(0.0249)	(0.0172)	(0.0069)
		$\tau_2 = 1 + \sqrt{c_n}$		
		$\cos\phi(u_{n,2}, e_2)$		
0.20	0.29	0.30	0.30	0.30
(0.1344)	(0.1558)	(0.1583)	(0.1565)	(0.1642)
		$\tau_3 = 1$		
		$\cos\phi(u_{n,3}, e_3)$		
0.05	0.05	0.06	0.06	0.06
(0.0409)	(0.0396)	(0.0428)	(0.0419)	(0.0419)
		$\tau_p = 1$		
		$\cos\phi(u_{n,p}, e_p)$		
0.06	0.06	0.06	0.06	0.06
(0.0427)	(0.0446)	(0.0423)	(0.0437)	(0.0410)

Note: $p = 200$, $n = 250$, and $c_n = 0.8$. The population covariance matrix is $\Sigma = \text{diag}(\tau_1, \tau_2, 1 \ldots, 1)$
τ_1 takes values in the set $\{2, 3, 4, 5, 10\}$, $\tau_2 = 1 + \sqrt{c_n}$ and $\tau_3 = \ldots = \tau_p = 1$. The number of Monte Carlo replications is $R = 1000$. The value of the cosine converging to 1 means that the angle is converging to 0. $e_1, \ldots, e_p$ are the basis vectors

Table 4.2 below confirms the bound results given in Proposition 4.3.1.

The results in Tables B.10 and B.11 in Appendix B provide further evidence for the cases $c_n = 0.5$ and $c_n = 0.2$.

The bound results for the cosine of the angle between the true $v_{n,i}$ and estimated eigenvectors $u_{n,i}$ can also be extended to the case of generalized spiked covariance model. In this case, Proposition 4.3.1 is as follows (see Yao et al. 2015, Section 11.3):

Proposition 4.3.4 *Let p, $n \to \infty$ such that $c_n = p/n \to c > 0$ and let $v_{n,i}$ denote the population eigenvector associated with τ_i, and let $u_{n,i}$ denote the eigenvector of sample covariance matrix $\widehat{S}_n$ associated with the eigenvalue $\lambda_{n,i}$. If $\psi'(\tau_i) > 0$, then*

$$|\langle u_{n,i}, v_{n,i} \rangle| \to \sqrt{\frac{\tau_i \psi'(\tau_i)}{\psi(\tau_i)}} \qquad a.s.$$

This result implies that if the population spike τ_i is fundamental, i.e., $\psi'(\tau_i) > 0$, the cosine of the angle between the true $v_{n,i}$ and estimated eigenvectors $u_{n,i}$ converges almost surely to a positive limit that is not equal to one. That is, asymptotically, space spanned by the estimated eigenvectors $u_{n,i}$ does not coincide with space spanned by the true eigenvectors $v_{n,i}$ in the generalized model.

The above results for sample eigenvectors in the spiked population model, according to Paul (2007), are based on the assumption of Gaussian data. However, the findings associated with sample eigenvectors under the generalized spiked population model established in Yao et al. (2015) require only i.i.d. data with the finite fourth moment, hence proving the universality of bound results in the non-null case. Recently, Bloemendal et al. (2016) extended the universality of bound results for the sample eigenvectors in the (generalized) spiked model under an assumption of the diagonal covariance matrix and the distribution of data having subexponential tails. Bloemendal et al. (2016) establish large deviation bounds for the sample eigenvectors and prove that the bound results similar to those in the Gaussian case (as in Paul 2007) hold in this setting. Furthermore, they derive the asymptotic Normal distribution of the sample eigenvectors corresponding to the large non-fundamental spikes in close vicinity to the transition threshold. An unexpected observation arising from that novel result is that, unlike the sample eigenvalues, sample eigenvectors contain information about the non-fundamental spikes.

The necessary consequence of the results presented above is that the use of PCA for dimension reduction cannot be justified in the high-dimensional setting—at least not in its standard form.

4.4 PCA Applications

As discussed earlier, the PCA is a useful multivariate statistics technique that aims at the dimension reduction of the observed data. It is particularly advantageous in applications with intrinsically low-dimensional data. From the discussion in the previous sections, we have already established that the underlying low-dimensional structure of the data is naturally linked to the eigenvalues and eigenvectors (see, Example 2.1.1 in Chap. 2). In what follows, we advance this discussion to demonstrate how the p-dimensional data can be generated so that it actually belongs to the one-dimensional space. The below example is according to Rigolett (2015).

Example 4.4.1 (The Design of Intrinsically Low-dimensional Random Data $Y_1, \dots, Y_n$) Let us fix a $p \times 1$ eigenvector v in a unit sphere $\mathbb{S}^{p-1}$,i.e., $v \in \mathbb{S}^{p-1}$. Furthermore, consider p-dimensional random variables $X_1, \dots, X_n$ that have a multivariate normal distribution $\mathcal{N}_p(0, I)$. Hence, each of the scalar variables $(v'X_1), \dots, (v'X_n)$ has a normal distribution $\mathcal{N}(0, 1)$. In turn, the random variables $(v'X_1)v, \dots, (v'X_n)v$ are p-dimensional. However, even though these vectors have p coordinates, they belong to the space spanned by only one eigenvector v. Intuitively, this means that the data points are clustered around the one-dimensional space. The random vectors $(v'X_1), \ldots, (v'X_n)$ are not observed; instead we analyze the variables $Y_1, \ldots, Y_n$ that are generated as follows:

$$Y_i = (v'X_i)v + \varepsilon_i,\ i = 1, \ldots, n,$$

where $\varepsilon_i \sim \mathcal{N}_p(0, \sigma^2 I)$ and are independent of X_i's. Thus, the covariance matrix of Y_i's generated as above is given as follows:

$$\Sigma = E(Y_i Y_i') = vv' + \sigma^2 I.$$

By rescaling, one can determine that $\Sigma = E(Y_i Y_i') = \tau_1 vv' + I$, where the spectrum of a population covariance matrix is $\text{spec}(\Sigma) = (\tau_1, 1, \ldots, 1)$, which is exactly the definition of a spiked covariance model according to Johnstone (2001), introduced earlier in Sect. 4.2.4.

The model from above example for the p-dimensional random variable Y_i is used by Johnstone and Lu (2004) in the high-dimensional setup ($n \to \infty,\ p = p(n) \to \infty,\ c_n = p/n \to c > 0$) to prove the inconsistency of the PCs. They call it "the single factor model" and define it as follows:

$$Y_i = b_i v_{n,1} + \varepsilon_i,\ i = 1, \ldots, n,$$

where Y_i is a column vector of observed data, $v_{n,1}$ is a latent factor in $\mathbb{R}^p$ (or the true eigenvector, which, however, depends on n in the high-dimensional setting), and ε_is are independent noise vectors with $\mathcal{N}_p(0, \sigma^2 I)$. The coefficients b_is are defined as i.i.d. Gaussian random effects with mean zero and unit variance, and are equivalent to $(v'X_i)$'s.

In the given framework, the sample eigenvector $u_{n,1}$ is shown to not be the consistent estimator of the population counterpart $v_{n,1}$. The idea of the proof is to show that in the limit the two vectors are not close to each other according to certain measure.

For the measure of the closeness between the sample eigenvector $u_{n,1}$ and the true eigenvector $v_{n,1}$, Johnstone and Lu (2004) chose $R(u_{n,1}, v_{n,1})$, which is the cosine of the angle between the two vectors, $R(u_{n,1}, v_{n,1}) = \cos\varphi(u_{n,1}, v_{n,1})$. Or equivalently, the inner product between the vectors after normalization to unit length:

$$R(u_{n,1}, v_{n,1}) = \langle u_{n,1}/\ ||u_{n,1}||, v_{n,1}/\ ||v_{n,1}||\rangle = u_{n,1}' v_{n,1}/\ ||u_{n,1}||\ ||v_{n,1}||.$$

This is a natural choice of measurement, and was used in the subsequent research (see, e.g., Paul 2007, Sect. 4.3).

Furthermore, it is assumed that the limiting "signal-to-noise ratio" in the model is as follows:

$$\lim_{n\to\infty} \frac{||v_{n,1}||^2}{\sigma^2} = \omega > 0.$$

Given all the assumptions and the specification of the model, the following bound result was obtained in the paper.

Proposition 4.4.1 *Assume that there are n observations drawn from the single factor model. Moreover, let* $n \to \infty$, $p = p(n) \to \infty$, $c_n = p/n \to c > 0$, *and* $||v_{n,1}||^2/\sigma^2 \to \omega > 0$. *Then,*

$$\lim_{n\to\infty} R^2(u_{n,1}, v_{n,1}) = R^2_\infty(\omega, c) = \frac{(\omega^2 - c)_+}{\omega^2 + c\omega} \; a.s.,$$

where $(x)_+ = \max\{x, 0\}$.

According to the above Proposition, if the limiting dimension-to-sample ratio $c > 0$ then the cosine of the angle between the two vectors in the limit is less than one, that is, $R_\infty(\omega, c) < 1$. This implies inconsistency of the sample eigenvector $u_{n,1}$ that corresponds to the largest eigenvalue under the high-dimensional asymptotics.

Furthermore, it is evident from the expression for $R_\infty(\omega, c)$ that the sample eigenvector $u_{n,1}$ is the consistent estimator of the true eigenvector $v_{n,1}$ if and only if $c_n = p/n \to 0$, which is the case under the traditional asymptotics.

The situation becomes worse if the limiting "signal-to-noise ratio" is smaller than the concentration ratio, $\omega^2 \leq c$. The latter means that there are insufficient observations to disentangle the true underlying factors from the noise. In this case, the two vectors, $u_{n,1}$ and $v_{n,1}$, are asymptotically orthogonal, and $u_{n,1}$ contains no information about $v_{n,1}$.

The results according to Paul (2007) extend the evidence by Johnstone and Lu (2004) to the case with k spike population eigenvalues, or equivalently, to the case of k sample eigenvectors. Furthermore, a detailed discussion on sample eigenvectors' inconsistency under spiked population model and a broad overview of PCA related phenomena are provided in Johnstone and Paul (2018).

To summarize, in the high-dimensional setup the noise does not average out in the factor models if there are too many dimensions p relative to the sample size n, which is why it is not possible to estimate the population eigenvectors consistently without any further restrictions imposed on the vectors, such as sparsity, for example. Paul and Aue (2014) provide an overview of the existing methods that are used to consistently estimate the eigenvectors, or principal components, under the sparsity assumption in the high-dimensional setting.[7]

[7] See Sect. 4.5 of the paper for sparse PCA, CCA, LDA, and covariance estimation.

4.5 Centered Sample Covariance Matrix and Substitution Principle

So far, we have studied the properties of the non-centered sample covariance matrix:

$$\widehat{S}_n^0 = \frac{1}{n}\sum_{i=1}^{n} Y_i' Y_i.$$

That is, it has been assumed that the data has mean equal to zero, $E(Y_i) = 0$. And the results discussed in previous sections rely on that assumption. However, in real-life applications, data has in general non-null mean, $E(Y_i) = \mu_Y$, and the covariance matrix is centered:

$$\widehat{S}_n = \frac{1}{n-1}\sum_{i=1}^{n}(Y_i - \bar{Y})'(Y_i - \bar{Y}), \text{ where } \bar{Y} = \frac{1}{n}\sum_{i=1}^{n} Y_i.$$

One can show that these two matrices are related to each other through the following representation:

$$\widehat{S}_n = \frac{1}{n-1}\sum_{i=1}^{n}(Y_i - \mu_Y)'(Y_i - \mu_Y) - \frac{n}{n-1}(\mu_Y - \bar{Y})'(\mu_Y - \bar{Y}),$$

where the first term is distributed as $\frac{n}{n-1}\widehat{S}_n^0$, whereas the second term is a random matrix of rank one.

This representation implies that the corresponding sample eigenvalues are different for the centered and non-centered sample covariance matrices (see, Section 3.5, Yao et al. 2015)[8]. However, the fact that sample eigenvalues for two matrices are different does not affect their limiting spectral distributions, as they have the same Marčenko–Pastur distribution when $c_n = p/n \to c$. In contrast, the linear spectral statistics of $\widehat{S}_n$ and $\widehat{S}_n^0$ do not share the same central limit theorem (CLT), as the asymptotic means are different (see, Pan 2014), and this further affects, for instance, the tests that are based on the spectral statistics, such as sphericity and partial sphericity tests. However, it was shown that the CLTs of the linear spectral statistics for the centered and non-centered matrices coincide if the *adjusted sample size* $N = n - 1$ is used in the centering parameter of the CLT for $\widehat{S}_n$. This result is referred to as *the substitution principle* and has been introduced in Zheng et al. (2015). Furthermore, this remarkable principle that provides a simple solution to the

---

[8] By Cauchy interlacing theorem: $\frac{n}{n-1}\lambda_1^0 \ge \lambda_1 \ge \frac{n}{n-1}\lambda_2^0 \ge \lambda_2 \ge \ldots \ge \frac{n}{n-1}\lambda_p^0 \ge \lambda_p$, where $\lambda_1^0, \ldots, \lambda_p^0$ are the sample eigenvalues of $\widehat{S}_n^0$, and $\lambda_1, \ldots, \lambda_p$ are, respectively, the sample eigenvalues of $\widehat{S}_n$.

centering problem holds both in the Gaussian and non-Gaussian cases, and hence, is universal.

4.6 Sphericity Test

As demonstrated in Sect. 3.5, the traditional sphericity test performs poorly in terms of its size when the dimension p is comparable to the sample size n, which is why Jiang and Yang (2013) introduced the correction of the traditional LRT under the high-dimensional asymptotics.

Let the classical LRT statistic be defined as $LRT_0 = \frac{|\widehat{S}_n|}{\left(\mathrm{tr}(\widehat{S}_n)/p\right)^p}$.[9] The observations are assumed to be multivariate normal distributed.

Proposition 4.6.1 *Let* $n \geq 3$, $p < n$ *and* $n \to \infty$, $p = p(n) \to \infty$ *such that* $c_n = p/n \to c \in (0, 1]$. *Then, under the null hypothesis* H_0 *the asymptotic distribution of* $\ln LRT_0$ *is given by:*

$$\frac{\ln LRT_0 - \mu_n}{\sigma_n} \xrightarrow{d} \mathcal{N}(0, 1),$$

$$\mu_n = -p - \left(n - p - \frac{3}{2}\right) \ln\left(1 - \frac{p}{n}\right), \quad \sigma_n^2 = -2\left[\frac{p}{n} + \ln\left(1 - \frac{p}{n}\right)\right].$$

Here it is necessary to note that the traditional LRT statistic for the sphericity is chi-square distributed under the standard asymptotics, whereas under the high-dimensional asymptotics the standardized version of this statistic is normally distributed.

As we can observe from Table 4.3, the corrected likelihood ratio test (CLRT) has an empirical size which is close to the nominal one even in the case when $p = 90$ and $n = 100$, while the traditional test (LRT) has an empirical size approaching one. The better performance in terms of the size of the test is, however, outweighed by the poor performance in terms of the empirical power, which gets worse with c getting close to one.

A theoretical explanation of the power deterioration when the concentration ratio p/n grows to one is provided in Section 5.1 of Forzani et al. (2017) for the case of partial sphericity test. Nevertheless, the same applies to the sphericity test when considering $k = 0$.

This test can be applied as well in case of $c > 1$. The correction of the test in this setup is suggested in Forzani et al. (2017). The technique used for the adjustment of the test statistic is discussed in the next section.

[9] Under the standard asymptotics the test statistic is defined in a slightly different way, namely, as $LRT_0^{n/2}$.

Table 4.3 Sphericity test. Corrected Likelihood Ratio test statistic. Empirical size and power

	c_n	Size $\widehat{\alpha}$		Power	
		CLRT	LRT	CLRT	LRT
n = 100, p = 5	0.05	0.0564	0.0521	0.6726	0.6505
n = 100, p = 30	0.3	0.0531	0.0620	0.8207	0.8384
n = 100, p = 60	0.6	0.0513	0.3276	0.7237	0.9633
n = 100, p = 90	0.9	0.0505	1.0000	0.4854	1.0000

Note: The empirical size $\widehat{\alpha}$ is estimated based on $R = 10000$ replications with $\Sigma = I$. The empirical power is estimated under the alternative hypothesis with $\Sigma = \text{diag}(1.65, \ldots, 1.65, 1, \ldots, 1)$, where the number 1.65 on the diagonal is equal to $[p/2]$. Based on Table 1 of Jiang and Yang (2013)

4.7 Partial Sphericity Test

In the high-dimensional setting, the test statistic is defined in the same way as under the standard asymptotics, that is:

$$LRT_k = \frac{\prod_{i=k+1}^{p} \lambda_{n,i}}{\left(\sum_{i=k+1}^{p} \lambda_{n,i}/(p-k)\right)^{(p-k)}}.$$

The results for the partial sphericity test can be derived for both cases $p < n$ and $p > n$. This was first pointed out by Srivastava (2006). Given the normality assumption imposed on the observed data, for $p < n$, under the null hypothesis H_0, $n\widehat{S}_n = nY_n'Y_n \sim \mathcal{W}_p(I_p, n)$.

However, one can build the likelihood ratio test for the *companion* sample covariance matrix defined as $\underline{\widehat{S}_n} = \frac{1}{n}Y_nY_n'$ (see, e.g., Sect. C.2), which has the size $n \times n$. Thus, for $p > n$, under the null hypothesis H_0, $n\underline{\widehat{S}_n} = nY_nY_n' \sim \mathcal{W}_n(I_n, p)$, i.e., Wishart-distributed with mean I_n and p degrees of freedom.

The non-zero eigenvalues for $\widehat{S}_n$ and $\underline{\widehat{S}_n}$ coincide. That is why, it is possible under H_0, in case $p > n$, modify the likelihood ratio test statistic as following for the sphericity test (which is special case of partial sphericity test when $k = 0$):

$$LRT_0 = \frac{|\underline{\widehat{S}_n}|}{\left(\text{tr}(\underline{\widehat{S}_n})/n\right)^{n}},$$

and in case of partial sphericity test, correspondingly, as:

$$LRT_k = \frac{\prod_{i=k+1}^{n} \lambda_{n,i}}{\left(\sum_{i=k+1}^{n} \lambda_{n,i}/(n-k)\right)^{(n-k)}}.$$

It is important to note here that in case $p > n$, the sample eigenvalues $\lambda_{n,n+1}, \ldots, \lambda_{n,p}$ are equal to zero, and hence, are omitted in the definitions of the test statistics.

The asymptotic distributions for LRT_0 and LRT_k in the high-dimensional setup are obtained under H_0 with n and p exchanging their roles. The asymptotic distributions are shown to be normal (see Forzani et al. 2017).

The following result relies on the condition that there exists such number $q_0 \ll \min(n, p)$ independent of p and n, that the hypothesis $H_k : \tau_{k+1} = \ldots = \tau_p$ is true for $k \leq q_0$.

Proposition 4.7.1 *Let $n\widehat{S}_n = nY_n'Y_n \sim \mathcal{W}_p(\Sigma_n, n)$ and let us assume the above condition fulfilled. Then, under the null hypothesis H_k that the true number of spikes is k and fixed,*

$$H_k : \tau_{k+1} = \ldots = \tau_p,$$

the asymptotic distribution of $\ln LRT_k$ (when $n, \ p \to \infty$ and $p/n \to c > 0$) is given by:

(i) Case $p < n$:

$$\frac{\ln LRT_k - \mu_{n,p,k}}{\sigma_{n,p,k}} \xrightarrow{d} \mathcal{N}(0, 1),$$

where

$$\mu_{n,p,k} = \tilde{\mu}_{n,p} + \ln A_{n,p,k} + \ln B_{n,p,k}, \quad \sigma^2_{n,p,k} = -2\left\{\frac{p-k}{n} + \ln\left(1 - \frac{p}{n}\right)\right\},$$

with

$$\tilde{\mu}_{n,p} = -p - \left(n - p - \frac{1}{2}\right)\ln\left(1 - \frac{p}{n}\right),$$

$$A_{n,p,k} = \frac{\prod_{i=1}^{k} \tau_i^{m_i}}{\prod_{i=i}^{k} \lambda_{n,i}},$$

$$B_{n,p,k} = \left(1 + \frac{\sum_{i=1}^{k} \lambda_{n,i} - \sum_{i=1}^{k} m_i \tau_i}{\sum_{i=k+1}^{p} \lambda_{n,i}}\right)^{p-k}.$$

The case $p > n + k$ is considered separately in Appendix C.4 for the sake of saving space. The asymptotic normality of the test statistic also holds in that case.

Here it is important to note that the random variable $\mu_{n,p,k}$ depends on the population eigenvalues τ_is. These values can be consistently estimated based on the bound results in Proposition 4.2.6 derived under the spiked population model assumption. The exact details are referred to Appendix C.4.

Table 4.4 Partial sphericity test. Corrected Likelihood Ratio test statistic. Empirical size

	c_n	Size $\widehat{\alpha}$			c_n	Size $\widehat{\alpha}$
		CLRT	LRT			CLRT
n = 100, p = 20	0.2	0.0028	0.0503	n = 400, p = 80	0.2	0.0418
n = 100, p = 30	0.3	0.0193	0.0596	n = 400, p = 120	0.3	0.0686
n = 100, p = 60	0.6	0.0724	0.2811	n = 400, p = 240	0.6	0.0933
n = 100, p = 90	0.9	0.0971	0.9999	n = 400, p = 360	0.9	0.0990

Note: The empirical size $\widehat{\alpha}$ is estimated based on $R = 10000$ replications. The distribution of random variables comprising the $n \times p$ data matrix Y_n is real Gaussian with mean zero and Σ, where $\Sigma = \text{diag}(7, 6, 5, 4, 1, \ldots, 1)$ is a spiked population covariance model with the number of spikes $k = 4$. The design is according to Forzani et al. (2017)

Corollary 4.7.1 *Under the conditions of Proposition 4.7.1, if the spike eigenvalues* $\tau_1, \ldots, \tau_k$ *are all greater than the threshold,* $\tau_i > \sigma^2 \left(1 + \sqrt{c}\right)$,

$$\frac{\ln LRT_k - \widehat{\mu_{n,p,k}}}{\sigma_{n,p,k}} \xrightarrow{d} \mathcal{N}(0, 1),$$

where $\widehat{\mu_{n,p,k}}$ *is obtained from* $\mu_{n,p,k}$ *by replacing* τ_i *with* $\tilde{\tau}_i$.

Table 4.4 presents the results of the Monte Carlo study for the empirical size of the test when testing the hypothesis that the number of spikes is equal to the true number $k = 4$.

The test for partial sphericity is performed sequentially. We consider the null hypothesis that $\Sigma = \sigma^2 I$. If this is rejected, we can test whether the $\min\{p-1, n-1\}$ smallest eigenvalues are equal, and so on. Thus, we test sequentially the null hypotheses for each $k \in (0, \ldots, q_0)$, where q_0 is as defined above.

The results given in Table 4.5 also confirm the power deterioration phenomenon that was observed earlier for the sphericity test. As we can see the likelihood ratio type of tests suffers from the power impairment, this is the reason why the information criteria type of tests were developed in this branch of literature recently. In what follows, we consider the tests by Kritchman and Nadler (2009) and Passemier and Yao (2014).

4.7.1 Kritchman and Nadler (2009) Test

The logic of the Kritchman and Nadler test is closely related to the traditional largest eigenvalue test of Roy (1953) under the standard asymptotics. However, it is

Table 4.5 Partial sphericity test. Corrected Likelihood Ratio test statistic. Empirical power

	c_n	$k = 1$	$k = 2$	$k = 3$	$\boldsymbol{k = 4}$	$k = 5$	$k = 6$	$k = 7$
n = 100, p = 20	0.2	0.00	0.00	0.03	0.97	0.00	0.00	0.00
n = 100, p = 30	0.3	0.00	0.00	0.11	0.88	0.01	0.00	0.00
n = 100, p = 60	0.6	0.00	0.03	0.54	0.40	0.02	0.01	0.00
n = 100, p = 90	0.9	0.04	0.36	0.42	0.13	0.02	0.01	0.01
n = 400, p = 80	0.2	0.00	0.00	0.00	0.96	0.02	0.00	0.00
n = 400, p = 120	0.3	0.00	0.00	0.01	0.93	0.02	0.01	0.00
n = 400, p = 240	0.6	0.00	0.00	0.41	0.55	0.02	0.01	0.00
n = 400, p = 360	0.9	0.01	0.27	0.50	0.17	0.02	0.01	0.01

Note: The values of k selected for the number of spikes over $R = 10000$ replications. The distribution of random variables comprising the $n \times p$ data matrix Y_n is real Gaussian with mean zero and Σ, where $\Sigma = \text{diag}(7, 6, 5, 4, 1, \ldots, 1)$ is a spiked population covariance model with number of spikes $k = 4$. The design is according to Forzani et al. (2017)

adjusted to the high-dimensional setup based on the fluctuation result for the largest eigenvalue.

In the null case (without any spikes) under an assumption that the variables $\{X_{ij}\}$ are i.i.d. standard Gaussian, the largest sample eigenvalue $\lambda_{n,1}$ is asymptotically Tracy–Widom distributed with parameters n and p (see Proposition 4.2.2):

$$\mathrm{P}\left(\frac{\lambda_{n,1}}{\sigma^2} < \mu_{np} + \sigma_{np} \cdot s\right) \xrightarrow{d} \mathcal{F}_1(s), \quad s > 0,$$

where μ_{np}, σ_{np} are as defined earlier.

By induction, this theoretical result can be extended further to the second largest sample eigenvalue $\lambda_{n,2}$ which would have as well Tracy–Widom distribution but with parameters n and $p - 1$. In the same manner, it can be shown that the k-th largest sample eigenvalue $\lambda_{n,k}$ is Tracy–Widom distributed with parameters n and $p - k$.

Under an assumption that the variance σ^2 is known, the Kritchman–Nadler criterion (detector) can be used to distinguish a sample eigenvalue $\lambda_{n,k}$ corresponding to the spike τ_k from a noise part at an asymptotic significance level α as following:

$$\lambda_{n,k} > \sigma^2 \left(\mu_{np} + \sigma_{n(p-k)} \cdot s(\alpha)\right),$$

Table 4.6 Partial sphericity test. Kritchman and Nadler statistic. Empirical power

	c_n	k = 1	k = 2	k = 3	**k = 4**	k = 5	k = 6	k = 7
n = 100, p = 20	0.2	0.00	0.00	0.00	0.96	0.04	0.00	0.00
n = 100, p = 30	0.3	0.00	0.00	0.00	0.95	0.05	0.00	0.00
n = 100, p = 60	0.6	0.00	0.00	0.00	0.91	0.08	0.00	0.00
n = 100, p = 90	0.9	0.00	0.00	0.00	0.89	0.10	0.00	0.00
n = 400, p = 80	0.2	0.00	0.00	0.00	0.96	0.04	0.00	0.00
n = 400, p = 120	0.3	0.00	0.00	0.00	0.95	0.05	0.00	0.00
n = 400, p = 240	0.6	0.00	0.00	0.00	0.93	0.07	0.00	0.00
n = 400, p = 360	0.9	0.00	0.00	0.00	0.91	0.09	0.00	0.00

Note: The values of k selected for the number of spikes over $R = 10000$ replications. The distribution of random variables comprising the $n \times p$ data matrix Y_n is real Gaussian with mean zero and Σ, where $\Sigma = \text{diag}(7, 6, 5, 4, 1, \dots, 1)$ is a spiked population covariance model with number of spikes $k = 4$. The design is according to Forzani et al. (2017)

where $s(\alpha)$ satisfies $\mathcal{F}_1(s(\alpha)) = 1 - \alpha$, i.e., is the corresponding quantile of the Tracy–Widom distribution [10] and

$$\sigma_{n(p-k)} = \left(\sqrt{n-1} + \sqrt{p-k}\right)\left(\frac{1}{\sqrt{n-1}} + \frac{1}{\sqrt{p-k}}\right)^{\frac{1}{3}}.$$

The Kritchman–Nadler test is performed sequentially on the nested hypotheses for the number of spikes $k = 1, 2, \dots, \min\{p, n\} - 1$.

For each value of k, if inequality $\lambda_{n,k} > \sigma^2 \left(\mu_{np} + \sigma_{n(p-k)} \cdot s(\alpha)\right)$ is satisfied, the null hypothesis is rejected and k is increased by 1.

The procedure stops once the null hypothesis is accepted, and the number of spikes is then estimated to be $\tilde{q}_n = k - 1$. Thus, the estimator of the number of spikes is defined as:

$$\tilde{q}_n = \min\left\{k \in \{1, \dots, q_0\} : \lambda_{n,k} < \hat{\sigma}^2 \left(\mu_{np} + \sigma_{n(p-k)} \cdot s(\alpha)\right)\right\} - 1,$$

where $\hat{\sigma}^2$ is a consistent estimator of the noise level σ^2 as defined in Appendix C.4.

The simulation results below present the performance of the Kritchman–Nadler test in the same setup as for the penalized version of the LRT (Table 4.6).

As we can see the test performs well, though there are no sophisticated tools employed to construct it. The test relies solely on the fluctuation result derived for the largest sample eigenvalue under the high-dimensional statistics. However, the

[10] The quantile of the Tracy–Widom distribution can be obtained using the R package RMTstat (available from the CRAN website at http://cran.r-project.org/web/packages/RMTstat/; see Johnstone et al. (2009) for documentation).

consistency of the test is established only in the "fixed p, large n" setting, which is one of the possible flaws of this procedure.

4.7.2 Passemier and Yao (2014) Test

For the illustrative purposes, the test in what follows is considered under Johnstone's spiked population model where all spikes have multiplicity one. The test is easily extended to the cases of generalized spiked population model and eigenvalues being not simple.

Given the bound results for the fundamental and non-fundamental spikes (see, e.g., Proposition 4.2.6), it is possible to construct an estimator for the number of spikes k based on the consecutive differences of the sample eigenvalues $\delta_{n,j} = \lambda_{n,j} - \lambda_{n,j+1}$, for $j \geq 1$. The idea is very similar to the one introduced by Onatski (2010).

It is shown that if $\tau_i > 1 + \sqrt{c}$, then $\lambda_{n,i} \rightarrow \tau_i \left(1 + \frac{c}{\tau_i - 1}\right)$ almost surely for $i = 1, \ldots, k$, and otherwise, if $1 < \tau_i \leq 1 + \sqrt{c}$, for sample eigenvalue holds almost sure convergence $\lambda_{n,i} \rightarrow (1 + \sqrt{c})^2$.

This implies that the difference $\delta_{n,j}$ almost surely tends to zero if $j > k$, and on the contrary, if $j \leq k$, then $\delta_{n,j}$ almost surely tends to a positive limit. Hence, the number of spikes k can be estimated as the index of the difference $\delta_{n,j}$ which first becomes small. Thus, the estimator is defined as follows:

$$\hat{q}_n = \min\left\{j \in \{1, \ldots, q_0\} : \delta_{n,j+1} < d_n\right\},$$

where $q_0 > k$ is a fixed number and d_n is a threshold defined below.

In case, where all the spikes have multiplicity one, it can be proved that the estimator is consistent, i.e., $\hat{q}_n \xrightarrow{p} k$, given that the threshold satisfies $d_n \rightarrow 0$, and $n^{2/3} d_n \rightarrow \infty$. Additionally some moment conditions are imposed on the random variables $\{X_{ij}\}$.

The practical implementation of the Passemier and Yao test depends on the choice of the threshold sequence d_n. The choice is provided to be of the form $d_n = Cn^{-2/3}\sqrt{2\log\log n}$, where the "tuning" parameter C can be selected based on a heuristic calibration procedure that is described in Appendix C.5.

The calibration of the threshold parameter in the Onatski's procedure is done using very different procedure. As pointed out by Onatski (2010), the estimator of the number of spikes can perform very poorly if the threshold is poorly calibrated. The results given in the Table 4.7 demonstrate this linkage.

Obviously, the chosen "tuning" parameter C is not optimal. This affects the empirical distribution of the number of chosen spikes negatively. However, with growing n, the performance improves significantly. Though, the test is still inferior to the alternative one.

Table 4.7 Partial sphericity test. Passemier and Yao statistic. Empirical power

	c_n	$k = 1$	$k = 2$	$k = 3$	$k = 4$	$k = 5$	$k = 6$	$k = 7$
n = 100, p = 20	0.2	0.06	0.06	0.00	0.71	0.12	0.01	0.00
n = 100, p = 30	0.3	0.07	0.09	0.00	0.76	0.03	0.00	0.00
n = 100, p = 60	0.6	0.09	0.10	0.00	0.72	0.03	0.00	0.00
n = 100, p = 90	0.9	0.12	0.12	0.02	0.65	0.02	0.00	0.00
n = 400, p = 80	0.2	0.03	0.02	0.00	0.92	0.00	0.00	0.00
n = 400, p = 120	0.3	0.03	0.02	0.00	0.91	0.00	0.00	0.00
n = 400, p = 240	0.6	0.05	0.04	0.00	0.86	0.00	0.00	0.00
n = 400, p = 360	0.9	0.05	0.04	0.00	0.86	0.00	0.00	0.00

Note: The values of k selected for the number of spikes over $R = 10000$ replications. The distribution of random variables comprising the $n \times p$ data matrix Y_n is real Gaussian with mean zero and Σ, where $\Sigma = \text{diag}(7, 6, 5, 4, 1, \ldots, 1)$ is a spiked population covariance model with number of spikes $k = 4$. The design is according to Forzani et al. (2017)

References

Akemann, G., Baik, J., & Di Francesco, P. (2011). *The Oxford handbook of random matrix theory*. Oxford University Press.

Anderson, T. W. (1963). Asymptotic theory for principal component analysis. *Annals of Mathematical Statistics, 34*(1), 122–148.

Bai, Z., & Silverstein, J. W. (2010). *Spectral analysis of large dimensional random matrices* (2nd ed.). New York: Springer.

Bai, Z., & Yao, J. (2008). Central limit theorems for eigenvalues in a spiked population model. *Annales de l'Institut Henri Poincaré, Probabilités et Statistiques, 44*(3), 447–474.

Bai, Z., & Yao, J. (2012). On sample eigenvalues in a generalized spiked population model. *Journal of Multivariate Analysis, 106*, 167–177.

Bai, Z. D., Liu, H. X., & Wong, W. K. (2011). Asymptotic properties of eigenmatrices of a large sample covariance matrix. *Annals of Applied Probability, 21*(5), 1994–2015.

Bai, Z. D., Miao, B. Q., & Pan, G. M. (2007). On asymptotics of eigenvectors of large sample covariance matrix. *Annals of Probability, 35*(4), 1532–1572.

Bai, Z. D., & Yin, Y. Q. (1993). Limit of the smallest eigenvalue of a large dimensional sample covariance matrix. *The Annals of Probability, 21*(7), 1275–1294.

Baik, J., Ben Arous, G., & Péché, S. (2005). Phase transition of the largest eigenvalue for nonnull complex sample covariance matrices. *The Annals of Probability, 33*(5), 1643–1697.

Baik, J., & Silverstein, J. W. (2006). Eigenvalues of large sample covariance matrices of spiked population models. *Journal of Multivariate Analysis, 97*(6), 1382–1408.

Bao, Z., Pan, G., & Zhou, W. (2015). Universality for the largest eigenvalue of sample covariance matrices with general population. *Annals of Statistics, 43*(1), 382–421.

Bloemendal, A., Knowles, A., Yau, H. T., & Yin, J. (2016). On the principal components of sample covariance matrices. *Probability Theory and Related Fields, 164*(1–2), 459–552.

Bloemendal, A., Knowles, A., Yau, H. T., & Yin, J. (2016). Direct shrinkage estimation of large dimensional precision matrix. *Journal of Multivariate Analysis, 146*, 223–236.

El Karoui, N. (2008). Spectrum estimation for large dimensional covariance matrices using random matrix theory. *Annals of Statistics, 36*(6), 2757–2790.

Fan, J., Liao, Y., & Liu, H. (2016). An overview of the estimation of large covariance and precision matrices. *The Econometrics Journal, 19*(1), C1–C32.

Fan, J., Lv, J., & Qi, L. (2011). Sparse high dimensional models in economics. *Annual Review of Economics, 3*, 291–317.

Feldheim, O. N., & Sodin, S. (2010). A universality result for the smallest eigenvalues of certain sample covariance matrices. *Geometric and Functional Analysis, 20*(1), 88–123.

Féral, D., & Péché, S. (2009). The largest eigenvalues of sample covariance matrices for a spiked population: Diagonal case. *Journal of Mathematical Physics, 50*(7), 073302.

Forzani, L., Gieco, A., & Tolmasky, C. (2017). Likelihood ratio test for partial sphericity in high and ultra-high dimensions. *Journal of Multivariate Analysis, 159*, 18–38.

Haff, L. R. (1980). Empirical Bayes estimation of the multivariate Normal covariance matrix. *The Annals of Statistics, 8*(3), 586–597.

Jiang, T., & Yang, F. (2013). Central limit theorems for classical likelihood ratio tests for high-dimensional normal distributions. *The Annals of Statistics, 41*(4), 2029–2074.

Johnstone, I. M. (2001). On the distribution of the largest eigenvalue in principal components analysis. *Annals of Statistics, 29*(2), 295–327.

Johnstone, I.M., & Lu, A.Y. (2004). *Sparse Principal Components Analysis*. Technical Report. Stanford University, Dept. of Statistics. Available: https://arxiv.org/abs/0901.4392

Johnstone, I.M., Ma, Z., Perry, P., & Shahram, M. (2009). *RMTstat: distributions, statistics and tests derived from random matrix theory* (R package version 0.2). https://cran.microsoft.com/snapshot/2014-09-08/web/packages/RMTstat/citation.html

Johnstone, I. M., & Paul, D. (2018). PCA in high dimensions: An orientation. *Proceedings of the IEEE, 106*(8), 1277–1292.

Kritchman, S., & Nadler, B. (2009). Non-parametric detection of the number of signals: Hypothesis testing and Random Matrix Theory. *IEEE Transactions on Signal Processing, 57*(10), 3930–3941.

Kritchman, S., & Nadler, B. (2012) Nonlinear shrinkage estimation of large-dimensional covariance matrices. *Annals of Statistics, 40*(2), 1024–1060.

Kritchman, S., & Nadler, B. (2015). Spectrum estimation: A unified framework for covariance matrix estimation and PCA in large dimensions. *Journal of Multivariate Analysis, 139*, 360–384.

Kritchman, S., & Nadler, B. (2017) Numerical implementation of the QuEST function. *Computational Statistics and Data Analysis, 115*, 199–223.

Lee, J. O., & Schnelli, K. (2016). Tracy–Widom distribution for the largest eigenvalue of real sample covariance matrices with general population. *The Annals of Applied Probability, 26*(6), 3786–3839.

Ma, Z. (2012). Accuracy of the Tracy–Widom limits for the extreme eigenvalues in white Wishart matrices. *Bernoulli, 18*(1), 322–359.

Marčenko, V. A., & Pastur, L. A. (1967). Distribution of eigenvalues for some sets of random matrices. *Mathematics of the USSR-Sbornik, 1*(4), 457–483.

Mehta, M. L. (1990). *Random matrices* (2nd ed.). Academic Press.

Mestre, X. (2008). Improved estimation of eigenvalues and eigenvectors of covariance matrices using their sample estimates. *IEEE Transactions on Information Theory, 54*(11), 5113–5129.

Onatski, A. (2010). Determining the number of factors from empirical distribution of eigenvalues. *Review of Economics and Statistics, 92*(4), 1004–1016.

Pan, G. (2014). Comparison between two types of large sample covariance matrices. *Annales de l'Institut Henri Poincaré, Probabilités et Statistiques, 50*(2), 655–677.

Passemier, D., & Yao, J. (2014). Estimation of the number of spikes, possibly equal, in the high-dimensional case. *Journal of Multivariate Analysis, 127*, 173–183.

Paul, D. (2007). Asymptotics of sample eigenstructure for a large dimensional spiked covariance model. *Statistica Sinica, 17*(4), 1617.

Paul, D., & Aue, A. (2014). Random matrix theory in statistics: A review. *Journal of Statistical Planning and Inference, 150*, 1–29.

Péché, S. (2009). Universality results for the largest eigenvalues of some sample covariance matrix ensembles. *Probability Theory and Related Fields, 143*(3–4), 481–516.

Peche, S., & Soshnikov, A. (2008). On the lower bound of the spectral norm of symmetric random matrices with independent entries. *Electronic Communications in Probability, 13*, 280–290.
Pillai, N. S., & Yin, J. (2014). Universality of covariance matrices. *Annals of Applied Probability, 24*(3), 935–1001.
Rigolett, P. (2015). *High-dimensional statistics*. Lecture notes. MIT.
Roy, S. (1953). On a heuristic method of test construction and its use in multivariate analysis. *The Annals of Mathematical Statistics, 24*(2), 220–238.
Silverstein, J., & Choi, S. (1995). Analysis of the limiting spectral distribution of large dimensional random matrices. *Journal of Multivariate Analysis, 54*(2), 295–309.
Silverstein, J. W. (1979). On the randomness of eigenvectors generated from networks with random topologies. *SIAM Journal on Applied Mathematics, 37*(2), 235–245.
Silverstein, J. W. (1981). Describing the behavior of eigenvectors of random matrices using sequences of measures on orthogonal groups. *SIAM Journal on Mathematical Analysis, 12*(2), 274–281.
Soshnikov, A. (2002). A note on universality of the distribution of the largest eigenvalues in certain sample covariance matrices. *Journal of Statistical Physics, 108*(5–6), 1033–1056.
Srivastava, M. S. (2006). Some tests criteria for the covariance matrix with fewer observations than the dimension. *Acta et Commentationes Universitatis Tartuensis de Mathematica, 10*, 77–93.
Stein, C. M. (1956). Inadmissibility of the usual estimator of the mean of a multivariate normal distribution. In *Proceedings of the Third Berkeley Symposium on Mathematical Statistics and Probability*. Berkeley: University of California Press.
Tao, T., & Vu, V. (2012). Random covariance matrices: Universality of local statistics of eigenvalues. *The Annals of Probability, 40*(3), 1285–1315.
Tracy, C. A., & Widom, H. (1996). On orthogonal and symplectic matrix ensembles. *Communications in Mathematical Physics, 177*(3), 727–754.
Wang, C., Pan, G., Tong, T., & Zhu, L. (2015). Shrinkage estimation of large dimensional precision matrix using random matrix theory. *Statistica Sinica, 25*(3), 993–1008.
Yao, J., Zheng, S., & Bai, Z. (2015). *Large sample covariance matrices and high-dimensional data analysis*. Cambridge University Press.
Yin, Y. Q., Bai, Z. D., & Krishnaiah, P. R. (1988). On the limit of the largest eigenvalue of the large dimensional sample covariance matrix. *Probability Theory and Related Fields, 78*(4), 509–521.
Zhang, M., Rubio, F., & Palomar, D. P. (2013). Improved calibration of high-dimensional precision matrices. *IEEE Transactions on Signal Processing, 61*(6), 1509–1519.
Zheng, S., Bai, Z., & Yao, J. (2015). Substitution principle for CLT of linear spectral statistics of high-dimensional sample covariance matrices with applications to hypothesis testing. *Annals of Statistics, 43*(2), 546–591.

Chapter 5
Summary and Outlook

This survey provides a holistic view of the properties of the sample covariance matrix and its eigendecomposition under two asymptotic regimes. The traditional regime is the basis of classical textbooks' statistics, while the high-dimensional regime better applies to Big Data context and closely approximates finite sample properties of standard estimators. Therefore, the objectives and contributions of this book are twofold.

The first is to collect all the theoretical results related to the sample covariance matrix and its respective eigenvalues and eigenvectors known in classical statistics and random matrix theory under one unifying and comprehensive topic: covariance matrix estimation. To the best of our knowledge, such a combination of results within one book has not yet been accomplished, as usually these results are discussed separately, without placing them into one perspective. Throughout the book, we emphasize that all the results are naturally interconnected and should be perceived as parts of common rather than independent findings. Such a common view is essential to understanding the current developments and open questions in modern covariance matrix related research directions.

The second objective is to integrate the two branches of literature: the results derived under the standard and high-dimensional asymptotics. These seemingly different branches, with the latter using more complex mathematical tools, are closely related, and one can draw insightful parallels. We demonstrate that a multitude of results in this new literature are a natural extension of the standard textbooks' findings, and therefore, the results derived under alternative asymptotics should also be introduced and made widely accessible to a broad audience. In addition, as the discussion indicates, many standard results flow into the new context with certain amendments (see also table in Sect. 5). Thus, the change of asymptotics does not overrule the previous results, instead adjusting them to the new realm and context.

A. Zagidullina, *High-Dimensional Covariance Matrix Estimation*,
SpringerBriefs in Applied Statistics and Econometrics,
https://doi.org/10.1007/978-3-030-80065-9_5

For example, in the standard "fixed p, large n" setup, if population eigenvalues are larger than one, the corresponding sample eigenvalues have the asymptotic Normal distribution. Moreover, these estimators are unbiased and consistent under traditional asymptotics (e.g., Anderson 1963). However, in the "large p, large n" setup, to obtain a similar result, one must pay a price for letting dimension p grow. First, the population eigenvalue should cross a certain threshold that depends on the concentration ratio, in other words, the limiting dimension p to sample size n ratio (see Baik et al. 2005). This condition, imposed on the population value, is akin to the one in the standard setup, but much stricter. Second, the corresponding sample eigenvalue converges almost surely to the biased population value, where the bias depends on the concentration ratio. Furthermore, the bias appears in the mean and variance of the asymptotic distribution that is Normal. This finding resembles the standard result under the traditional asymptotics but with corresponding adjustments for growing p in the high-dimensional setting. In addition, a finite sample bias derived by Lawley (1956) for the sample eigenvalues under standard asymptotics appears in the limiting mean value of the sample eigenvalue estimator under the high-dimensional asymptotics, though with a correction factor (see Mestre 2008).

Another closely related example that demonstrates the similarities of the findings under the two asymptotic regimes and provides further insights into the applications is concerned with the covariance matrix eigenvectors. In the "large p, large n" setting, the sample eigenvectors (corresponding to the population eigenvalues that cross the threshold) are asymptotically multivariate Gaussian. This result is also reminiscent of the standard result in multivariate analysis. However, in the limit, the sample eigenvectors do not converge to the population values. This inconsistency result is a theoretical justification for why the PCA cannot be used in its standard form in the high-dimensional framework (see Johnstone & Paul 2018), and additional structures should be imposed on the population vectors to achieve consistency in estimation.

Thus, when examining the results under high-dimensional asymptotics, one can associate these results with the traditional ones and understand how the relative magnitude of dimension p and sample size n affects the consistency property of the sample estimators. As far as we know, such analogies and comparative study have not yet been presented. Therefore, to contribute to the existing literature, this book builds a bridge between the two statistical frameworks and provides a broad overview that unifies and structures seemingly unrelated results.

The survey focuses on the covariance matrix and demonstrates the link between the two asymptotic frameworks with this specific example. However, the extension of classical results to the high-dimensional setting using RMT techniques can be established for many statistical tools.

For example, the RMT can be used to construct the linear spectral statistics and derive the respective limiting distributions for the high-dimensional inference. The linear spectral statistic based on the largest sample eigenvalue was first introduced in Johnstone (2001) to test the hypothesis that the observations are independently and identically distributed in high dimensions. Given that the above hypothesis is too restrictive in many applications, recent research into the high-dimensional inference turned to more general independence tests. In Pan et al. (2014), the spectral statistic was constructed based on the sample covariance matrix to test the independence of the large number of high-dimensional random vectors with a general structure. In particular, these vectors can represent the observations from multivariate time series or the linear combinations of independent variables. Testing for the independence of the random vector's components was developed in Gao et al. (2017) based on the spectral statistic of the sample correlation matrix. Furthermore, the authors demonstrate that this test can also be applied in the factor model context to infer the equivalence of factor loadings. The test for the hypothesis that the sub-vectors of a high-dimensional random vector are independent, and hence the covariance matrix has a block-diagonal structure, was investigated in Bodnar et al. (2019). Moreover, Yang et al. (2020) extended the previous work by introducing the sphericity test for high-dimensional covariance matrices under a generalized elliptical model that accommodates the heavy tails, heteroscedasticity, and asymmetry observed in the real data. These tests impose moment restrictions on the data, but the linear spectral statistics can also be designed in the non-parametric setting. For instance, the distribution-free statistic to test the independence of the random vector's components was proposed in Bao et al. (2015) based on the non-parametric Spearman's rank correlation matrix. The above references represent only part of the literature dedicated to the specific test type. However, high-dimensional inference built upon the linear spectral statistics is an active area of modern research. Thus, a growing body of literature is concerned with multivariate statistical tests and their amendments in the high-dimensional setting.

Moreover, the RMT tools introduced in this survey are used to construct the shrinkage type of estimators for the covariance and precision matrices in the absence of any distributional or structural assumptions. These estimators remediate the sample covariance (precision) matrix's deficiencies in the high-dimensional setting by shrinking the sample estimator to some known target, thus reducing the variability of the estimator at the cost of a bias. The RMT in this context is applied to derive the optimal and consistent shrinkage parameters (or intensities) and/or the non-linear shrinkage transformations of the sample eigenvalues. In this branch of literature, Ledoit and Wolf (2004) were first to introduce the covariance matrix estimator, which is a linear combination of the sample covariance matrix with the identity matrix. They proved that under the high-dimensional asymptotics, the estimated shrinkage intensities are consistent, and the covariance estimator is optimal with respect to the squared loss function. Recently, Bodnar et al. (2014) extended this work to a case in which the target covariance matrix is an arbitrary symmetric positive definite matrix with uniformly bounded trace norm. Using RMT, they revealed that, in this general case, the optimal shrinkage parameters are also consistently estimated, and the resulting covariance estimator obeys almost surely the smallest Frobenius norm. Alternatively, Ledoit and Wolf (2015) developed the covariance matrix estimator, which builds upon the non-linear transformation of sample eigenvalues and, hence, is a non-linear shrinkage estimator of the covariance matrix. Results similar to those above were obtained for this non-linear shrinkage estimator as well. The shrinkage estimators of precision matrices in the high-dimensional setting are constructed similarly. They are based either on the inverse of the corresponding shrinkage covariance estimator or on the direct shrinkage of the sample covariance matrix's inverse toward some target. The first type of estimator was developed and analyzed from the RMT perspective in Wang et al. (2015). The second type of estimator was introduced by Bodnar et al. (2016) and studied for a case in which the target matrix is an arbitrary non-random matrix with a uniformly bounded normalized trace norm. Although this is an appealing research direction within the covariance matrix estimation field, few results are known. However, with recent developments in the RMT, we believe this area will advance in the future.

Thus, there are numerous research directions in which the enhancement of RMT may be beneficial in dealing with high-dimensional data that are not covered in this book. Furthermore, the key RMT findings we have considered in this book are derived under simplifying assumptions imposed on the data matrix, but recent developments in this field allow for adjusting these results using less restrictive requirements. Therefore, we want to highlight this generic direction for improvement, which is concerned with the generalization of assumptions imposed on the data structures under the high-dimensional asymptotics.

Many RMT results are developed in the setting of i.i.d. observations with the Normality assumption imposed on the data matrix entries, but typical economic or financial data depends on time. Thus, an extension of the current theory on the eigenvalues of Wishart-type matrices, in which the columns of the data matrix are considered a realization of a high-dimensional multivariate time series, is crucial for modern econometrics and finance. The recent results by Yao (2012) and Pfaffel and Schlemm (2011), among others, are concerned with extensions of the Marčenko–Pastur result to linear stationary time series with short-range or long-range dependence. Furthermore, a study of the sample autocovariance matrix properties under high-dimensional asymptotics is important for the understanding of how the dimensionality p affects the inference of the high-dimensional times series. Li et al. (2015), Liu et al. (2015), Wang and Yao (2016) and Wang et al. (2017), among others, analyzed the behavior of the empirical spectral distribution of autocovariance matrices under certain conditions imposed on the data matrix (also allowing for temporal dependence in the data). An investigation of the spectrum properties for the sample auto-cross covariance matrices is closely related to the areas of research described above. These matrices are based on the data generated from the dynamic factor models and are essential for the estimation of a number of factors and their lags from the high-dimensional time series arising in diverse fields, such as macroeconomics, finance, and signal processing. Recently, Jin et al. (2014) derived the limiting spectral distribution for this type of matrix, and Wang et al. (2015) established the limits of the largest and smallest eigenvalues. Additionally, RMT enhancement can be beneficial when studying the properties of the covariance matrices stemming from high-frequency data and data with missing observations (e.g., Jurczak & Rohde 2017). Considerable effort has been devoted to generalizing the results of RMT beyond independence or linear dependence structure in the columns of the data, for example, by considering the elliptical distributions that allow for non-linear dependence and induce tail dependence, which is crucial in an analysis of financial markets (e.g., El Karoui 2009 and Hu et al. 2019). Moreover, the major thread of research in modern RMT is concerned with establishing the universality that refers to the phenomenon that the asymptotic behavior of eigenvalue–eigenvector statistics does not depend on the distribution of data (see discussion in Chap. 4, Paul and Aue 2014 and Akemann et al. 2011).

Thus, there are numerous remaining challenges and open questions in the estimation of covariance matrices and their spectral statistics, eigenvectors. Currently, RMT is an active area of contemporary research because its framework provides powerful tools to explore properties of high-dimensional data and facilitates the analysis of phenomena observed in Big Data. Hence, RMT findings will have a marked impact on econometric theory and practice, from statistical modeling to fundamental understanding of problems in many applied fields of research. Therefore, the discussion in this book contributes to the existing literature by providing a timely overview of key results in this highly technical area, guiding readers through the recent developments and facilitating the comprehension of other related topics beyond the scope of this survey.

Comparative Study of Statistical Properties Under Two Asymptotic Regimes. Traditional Estimators and LRT Statistics

Sample Covariance	
Traditional Asymptotics	High-dimensional Asymptotics
• $\Sigma = I$ $\sqrt{n}(\hat{S}_n - I) \xrightarrow{d} G(W)$ The distribution of the Wigner matrix • Σ is arbitrary $B_n = \sqrt{n}(\hat{S}_n - \Sigma) \xrightarrow{d} \mathcal{N}(0, \Sigma_B)$	• $\Sigma = I$ no result • Σ is arbitrary no result

Sample Eigenvalues	
Traditional Asymptotics	High-dimensional Asymptotics
• $\Sigma = I$ $\sqrt{n}(\Lambda_n - I) \xrightarrow{d} G(l_1^w, l_2^w, \ldots, l_p^w)$ The distribution of the Wigner matrix eigenvalues	Null Case: • $\Sigma = I$ $f_{c,\sigma^2}(x) = \frac{1}{2\pi c\sigma^2} \frac{\sqrt{b_c - x}\sqrt{x - a_c}}{x} \mathbb{1}\{a_c \le x \le b_c\}$ The pdf of the Marčenko-Pastur distribution
• Σ is arbitrary If $\tau_1 > \ldots > \tau_p > 1$ $\sqrt{n}(\lambda_{n,i} - \tau_i) \xrightarrow{d} \mathcal{N}(0, 2\tau_i^2), \quad i = 1, \ldots, p.$	• Σ is arbitrary $f_{c,H}(x) = \frac{1}{\pi} \lim_{\varepsilon \to 0} \mathrm{Im}[m(x + i\varepsilon)],$ $m(z) = \int \frac{1}{\tau(1 - c - czm(z)) - z} dH(\tau), z \in \mathbb{C}^+.$ The pdf of generalized Marčenko-Pastur distribution Non-null case:
• Σ is arbitrary If $\tau_1 > \ldots > \tau_p > 1$ $\sqrt{n}(\lambda_{n,i} - \tau_i) \xrightarrow{d} \mathcal{N}(0, 2\tau_i^2), \quad i = 1, \ldots, p.$	Johnstone's spiked model • $\Sigma = \mathrm{diag}(\tau_1, \ldots, \tau_k, 1, \ldots, 1)$ If $\tau_i > 1 + \sqrt{c}$ $\sqrt{n}(\lambda_{n,i} - \rho(\tau_i)) \xrightarrow{d} \mathcal{N}_k(0, \sigma^2(\tau_i)), \quad i = 1, \ldots, k$ $\rho(\tau) = \tau\left(1 + \frac{c}{\tau - 1}\right), \sigma^2(\tau) = 2\tau^2\left(1 - \frac{c}{(\tau - 1)^2}\right).$
• Σ is arbitrary If $\tau_1 > \ldots > \tau_p > 1$ $\sqrt{n}(\lambda_{n,i} - \tau_i) \xrightarrow{d} \mathcal{N}(0, 2\tau_i^2), \quad i = 1, \ldots, p.$	Generalized spiked model • $\mathrm{spec}(\Sigma) = \mathrm{diag}(\tau_1, \ldots, \tau_k, \beta_{n,1}, \ldots, \beta_{n,p-k})$ If $\psi'(\tau_i) > 0$ $\sqrt{n}(\lambda_{n,i} - \psi(\tau_i)) \xrightarrow{d} \mathcal{N}_k(0, \sigma^2(\tau_i)), \quad i = 1, \ldots, k$ $\psi(\tau) = \psi_{c,H}(\tau) = \tau + c\int \frac{t\tau}{\tau - t} dH(t),$ $\sigma^2(\tau_i) = 2\tau_i^2\psi'(\tau_i).$

Sample Eigenvectors

Traditional Asymptotics	High-dimensional Asymptotics
	Null Case:
• $\Sigma = I$	• $\Sigma = I,\ \Sigma$ is arbitrary
the conditional Haar distribution under the Gaussian assumption	the conditional Haar distribution under the Gaussian assumption
no result under the non-Gaussianity	the *asymptotic* conditional Haar distribution under the non-Gaussianity assumption
	Non-null case:
	Johnstone's spiked model
• Σ is arbitrary	• $\Sigma = \mathrm{diag}(\tau_1, \ldots, \tau_k, 1, \ldots, 1)$
If $\tau_1 > \ldots > \tau_p > 1$	If $\tau_i > 1 + \sqrt{c}$
$\sqrt{n}(u_{n,i} - v_i) \xrightarrow{d} \mathcal{N}(0, \Sigma(\tau_i)),$	$\sqrt{n}\left(\frac{u_{n,iA}}{\Vert u_{n,iA}\Vert} - e_i\right) \xrightarrow{d} \mathcal{N}_k(0, \Sigma(\tau_i)),$
$\Sigma(\tau_i) = \tau_i \sum_{1 \le k \ne i \le p} \frac{\tau_k}{(\tau_k - \tau_i)^2} v_k v_k'.$	$\Sigma(\tau_i) = \frac{\tau_i}{1 - \frac{c}{(\tau_i - 1)^2}} \cdot \sum_{1 \le l \ne i \le k} \frac{\tau_l}{(\tau_l - \tau_i)^2} e_l e_l'.$
	$\frac{u_{n,iB}}{\Vert u_{n,iB}\Vert}$ is distributed uniformly on the unit sphere $\mathbb{S}^{p-k-1}$

Sphericity test

Traditional Asymptotics	High-dimensional Asymptotics
	$n \ge 3, \quad p/n \to c \in (0, 1]$ for $n \to \infty$
• $-2p \ln LRT_0 \xrightarrow{d} \chi^2_{(df)}$	• $\frac{\ln LRT_0 - \mu_n}{\sigma_n} \xrightarrow{d} \mathcal{N}(0, 1)$
	$\mu_n = -p - \left(n - p - \frac{3}{2}\right) \ln\left(1 - \frac{p}{n}\right)$
	$\sigma_n^2 = -2\left[\frac{p}{n} + \ln\left(1 - \frac{p}{n}\right)\right]$

Partial sphercity test

Traditional Asymptotics	High-dimensional Asymptotics
• $-p \ln LRT_k \xrightarrow{d} \chi^2_{(df)}$	• $\frac{\ln LRT_k - \widehat{\mu_{n,p,k}}}{\sigma_{n,p,k}} \xrightarrow{d} \mathcal{N}(0, 1)$

References

Akemann, G., Baik, J., & Di Francesco, P. (2011). *The Oxford handbook of random matrix theory*. Oxford University Press.

Anderson, T. W. (1963). Asymptotic theory for principal component analysis. *Annals of Mathematical Statistics, 34*(1), 122–148.

Baik, J., Ben Arous, G., & Péché, S. (2005). Phase transition of the largest eigenvalue for nonnull complex sample covariance matrices. *The Annals of Probability, 33*(5), 1643–1697.

Bao, Z., Lin, L.-C., Pan, G., & Zhou, W. (2015). Spectral statistics of large dimensional Spearman's rank correlation matrix and its application. *Annals of Statistics, 43*, 2588–2623.

Bodnar, T., Dette, H., & Parolya, N. (2019). Testing for independence of large dimensional vectors. *Annals of Statistics, 47*(5), 2977–3008.

Bodnar, T., Gupta, A. K., & Parolya, N. (2014). On the strong convergence of the optimal linear shrinkage estimator for large dimensional covariance matrix. *Journal of Multivariate Analysis, 132*, 215–228.

Bodnar, T., Gupta, A. K., & Parolya, N. (2016). Direct shrinkage estimation of large dimensional precision matrix. *Journal of Multivariate Analysis, 146*, 223–236.

Bodnar, T., Gupta, A. K., & Parolya, N. (2009). Concentration of measure and spectra of random matrices: Applications to correlation matrices, elliptical distributions and beyond. *The Annals of Applied Probability, 19*(6), 2362–2405.

Gao, J., Han, X., Pan, G., & Yang, Y. (2017). High dimensional correlation matrices: The central limit theorem and its applications. *Journal of the Royal Statistical Society. Series B: Statistical Methodology, 79*(3), 677–693.

Hu, J., Li, W., Liu, Z., & Zhou, W. (2019). High-dimensional covariance matrices in elliptical distributions with application to spherical test. *The Annals of Statistics, 47*(1), 527–555.

Jin, B., Wang, C., Bai, Z. D., Nair, K. K., & Harding, M. (2014). Limiting spectral distribution of a symmetrized auto-cross covariance matrix. *Annals of Applied Probability, 24*(3), 1199–1225.

Johnstone, I. M. (2001). On the distribution of the largest eigenvalue in principal components analysis. *Annals of Statistics, 29*(2), 295–327.

Johnstone, I. M., & Paul, D. (2018). PCA in high dimensions: An orientation. *Proceedings of the IEEE, 106*(8), 1277–1292.

Jurczak, K., & Rohde, A. (2017). Spectral analysis of high-dimensional sample covariance matrices with missing observations. *Bernoulli, 23*(4A), 2466–2532.

Lawley, D. N. (1956). Tests of significance for the latent roots of covariance and correlation matrices. *Biometrika, 43*(1–2), 128–136.

Ledoit, O., & Wolf, M. (2004). A well-conditioned estimator for large-dimensional covariance matrices. *Journal of Multivariate Analysis, 88*(2), 365–411.

Ledoit, O., & Wolf, M. (2015). Spectrum estimation: A unified framework for covariance matrix estimation and PCA in large dimensions. *Journal of Multivariate Analysis, 139*, 360–384.

Li, Z., Pan, G., & Yao, J. (2015). On singular value distribution of large-dimensional autocovariance matrices. *Journal of Multivariate Analysis, 137*, 119–140.

Liu, H., Aue, A., & Paul, D. (2015). On the Marčenko–Pastur law for linear time series. *Annals of Statistics, 43*(2), 675–712.

Mestre, X. (2008). Improved estimation of eigenvalues and eigenvectors of covariance matrices using their sample estimates. *IEEE Transactions on Information Theory, 54*(11), 5113–5129.

Pan, G., Gao, J., & Yang, Y. (2014). Testing independence among a large number of high-dimensional random vectors. *Journal of the American Statistical Association, 109*(506), 600–612.

Paul, D., & Aue, A. (2014). Random matrix theory in statistics: A review. *Journal of Statistical Planning and Inference, 150*, 1–29.

Pfaffel, O., & Schlemm, E. (2011). Eigenvalue distribution of large sample covariance matrices of linear processes. *Probability and Mathematical Statistics, 31*(2), 313–329.

Wang, C., Jin, B., Bai, Z. D., Nair, K. K., & Harding, M. (2015). Strong limit of the extreme eigenvalues of a symmetrized auto-cross covariance matrix. *Annals of Applied Probability, 25*(6), 3624–3683.

Wang, C., Pan, G., Tong, T., & Zhu, L. (2015). Shrinkage estimation of large dimensional precision matrix using random matrix theory. *Statistica Sinica, 25*(3), 993–1008.

Wang, L., Aue, A., & Paul, D. (2017). Spectral analysis of sample autocovariance matrices of a class of linear time series in moderately high dimensions. *Bernoulli, 23*(4A), 2181–2209.

Wang, Q., & Yao, J. (2016). Moment approach for singular values distribution of a large auto-covariance matrix. *Annales de l'Institut Henri Poincaré, Probabilités et Statistiques, 52*(4), 1641–1666.

Yang, X., Zheng, X., & Chen, J. (2021). Testing high-dimensional covariance matrices under the elliptical distribution and beyond. *Journal of Econometrics, 221*(2), 409–423.

Yao, J. (2012). A note on a Marčenko-Pastur type theorem for time series. *Statistics & Probability Letter, 82*(1), 22–28.

Appendix A
Graphs

See Figs. A.1, A.2, A.3, A.4, A.5, A.6, A.7, A.8, A.9, and A.10.

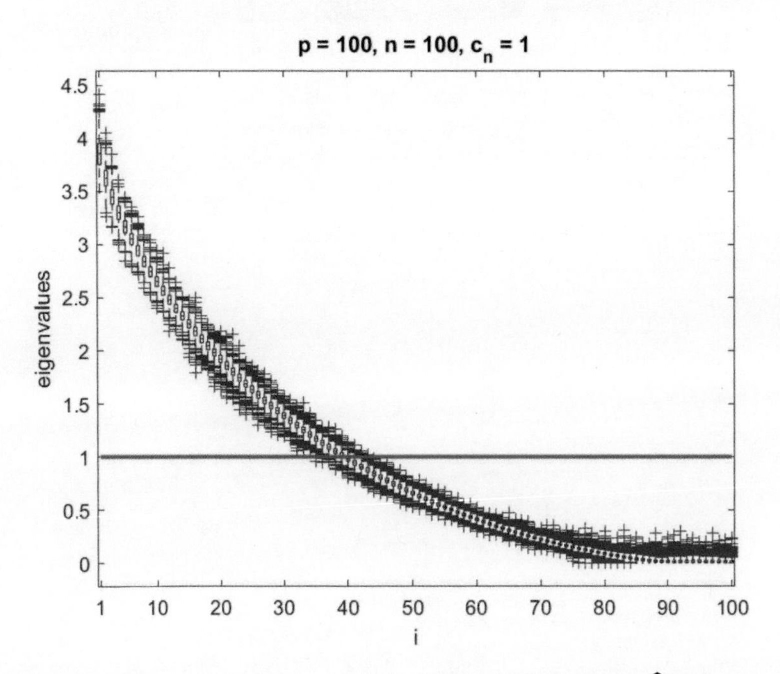

Fig. A.1 The boxplots for the eigenvalues of the sample covariance matrix $\widehat{S}_n$ over $R = 1000$ Monte Carlo replications. The blue thick line indicates the unit population eigenvalues. The distribution of random variables comprising the $n \times p$ data matrix Y_n is real Gaussian with mean zero and $\Sigma = I$

A. Zagidullina, *High-Dimensional Covariance Matrix Estimation*,
SpringerBriefs in Applied Statistics and Econometrics,
https://doi.org/10.1007/978-3-030-80065-9

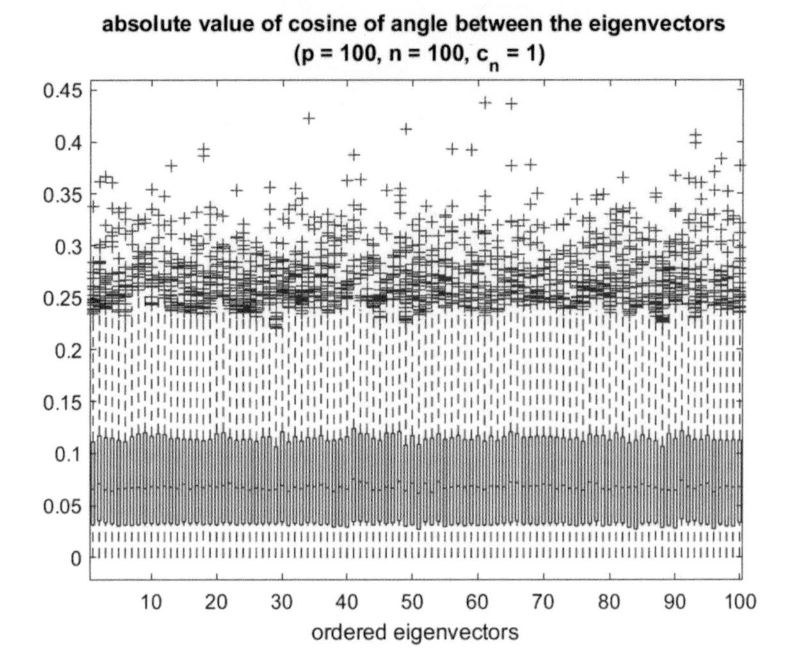

Fig. A.2 The boxplots of $\cos\phi(u_{n,i}, v_i)$, the cosine of the angle between the sample eigenvector and the true one over $R = 1000$ Monte Carlo replications. The distribution of random variables comprising the $n \times p$ data matrix Y_n is real Gaussian with mean zero and $\Sigma = I$

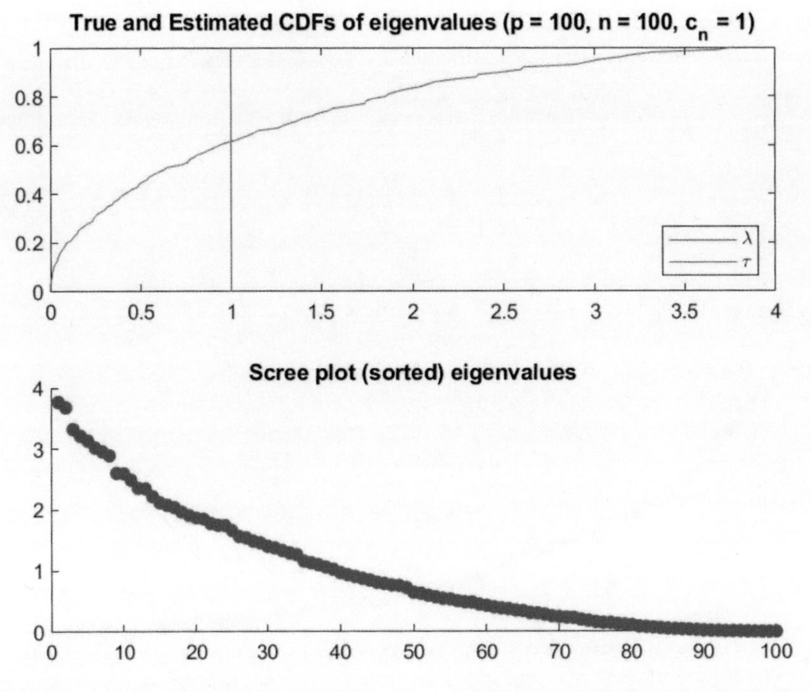

Fig. A.3 Case $\Sigma = I$. Figure is generated using one realization of the data. The distribution of random variables comprising the $n \times p$ data matrix Y_n is real Gaussian with mean zero and $\Sigma = I$. The c.d.f. of the true eigenvalues is a point mass at 1, δ_1. Based on Figure 1 of El Karoui (2008)

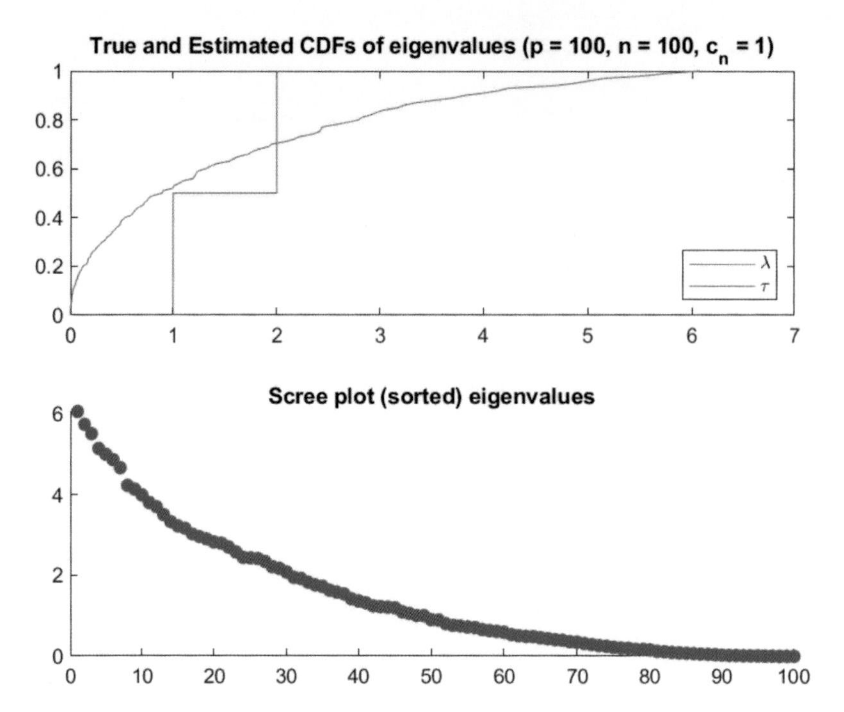

Fig. A.4 Case $\Sigma = I_{(p/2)} + 2 \cdot I_{(p/2)}$. Figure is generated using one realization of the data. The distribution of random variables comprising the $n \times p$ data matrix Y_n is real Gaussian with mean zero and $\Sigma = I_{(p/2)} + 2 \cdot I_{(p/2)}$. The c.d.f. of the true eigenvalues are point masses at 1 and at 2, i.e., $0.5\delta_1 + 0.5\delta_2$. Based on Figure 3 of El Karoui (2008)

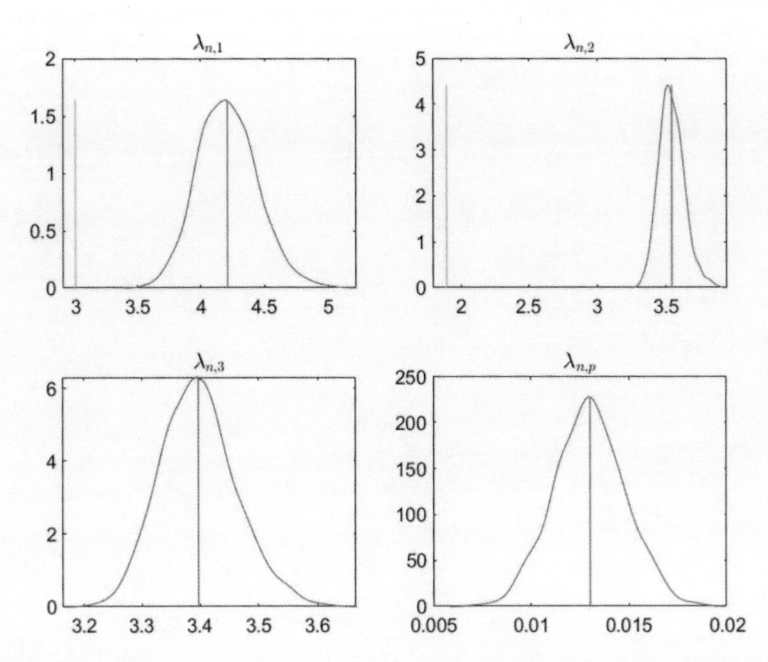

Fig. A.5 $p = 200$, $n = 250$, $c_n = 0.8$. The population covariance matrix is $\Sigma = \text{diag}(\tau_1, \tau_2, 1 \ldots, 1)$. $\tau_1 = 3$, $\tau_2 = 1 + \sqrt{c_n}$ and $\tau_3 = \ldots = \tau_p = 1$. The number of Monte Carlo replications is $R = 1000$. The densities for $\lambda_{n,1}$, $\lambda_{n,2}$, $\lambda_{n,3}$, and $\lambda_{n,p}$ are estimated using the "Normal" kernel density

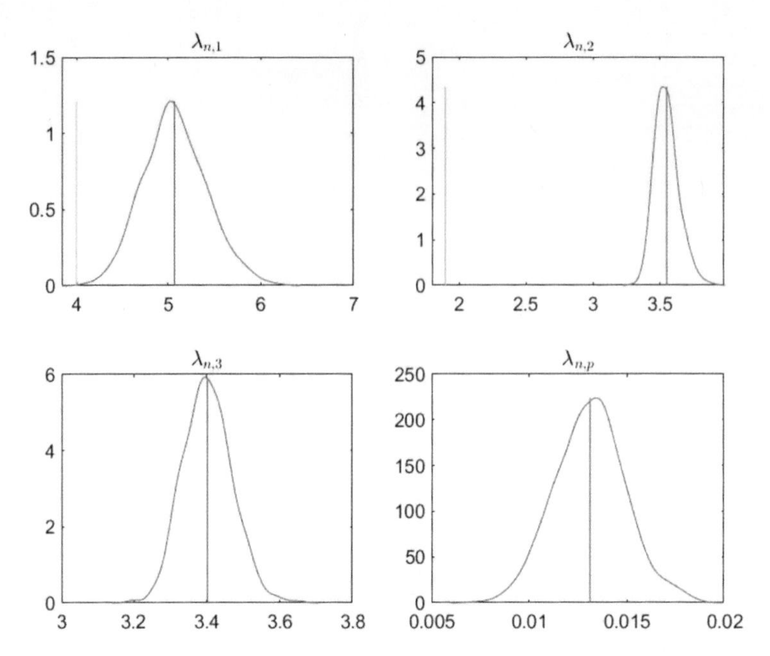

Fig. A.6 $p = 200$, $n = 250$, $c_n = 0.8$. The population covariance matrix is $\Sigma = \text{diag}(\tau_1, \tau_2, 1 \ldots, 1)$. $\tau_1 = 4$, $\tau_2 = 1 + \sqrt{c_n}$, and $\tau_3 = \ldots = \tau_p = 1$. The number of Monte Carlo replications is $R = 1000$. The densities for $\lambda_{n,1}$, $\lambda_{n,2}$, $\lambda_{n,3}$, and $\lambda_{n,p}$ are estimated using the "Normal" kernel density

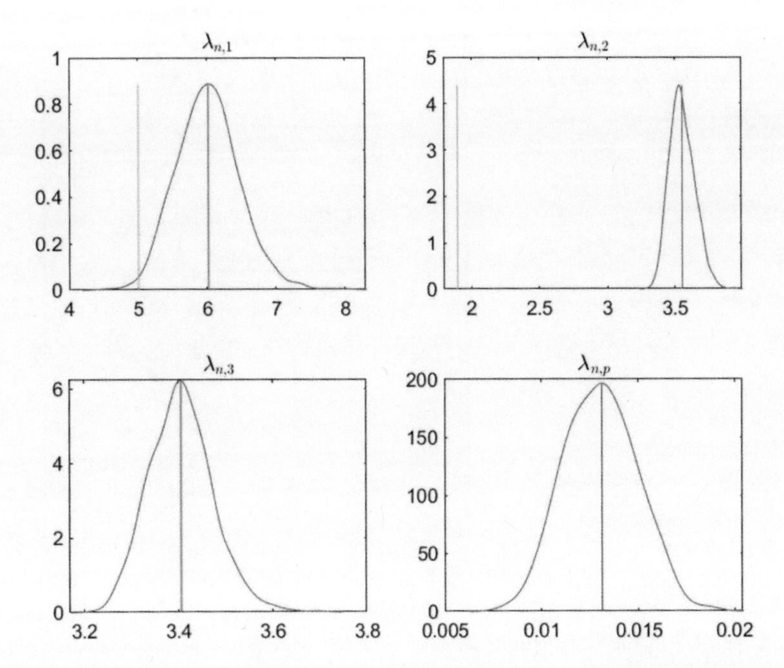

Fig. A.7 $p = 200$, $n = 250$, $c_n = 0.8$. The population covariance matrix is $\Sigma = \text{diag}(\tau_1, \tau_2, 1 \ldots, 1)$. $\tau_1 = 5$, $\tau_2 = 1 + \sqrt{c_n}$, and $\tau_3 = \ldots = \tau_p = 1$. The number of Monte Carlo replications is $R = 1000$. The densities for $\lambda_{n,1}$, $\lambda_{n,2}$, $\lambda_{n,3}$, and $\lambda_{n,p}$ are estimated using the "Normal" kernel density

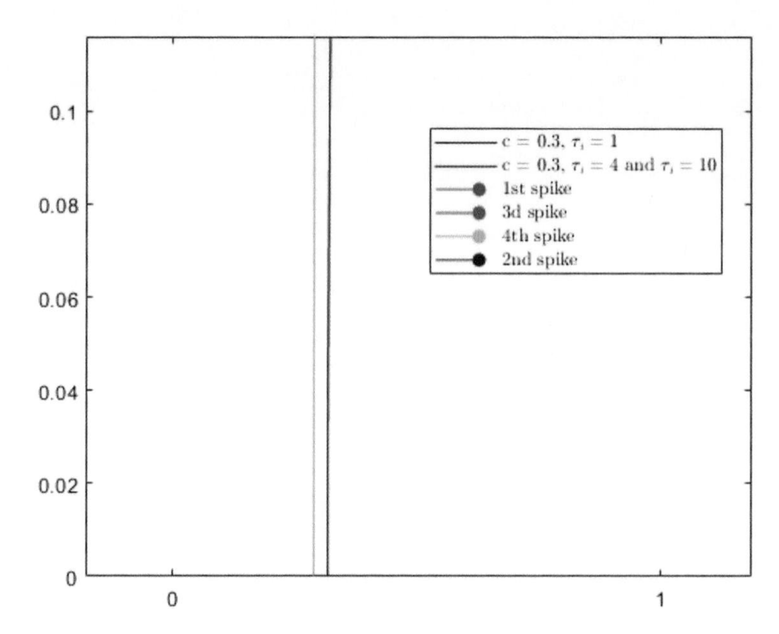

Fig. A.8 The density function for the limiting spectral distribution $F_{c,H}$ of the sample eigenvalues, where $c = 0.3$ and H is uniform on the set $\{1, 4, 10\}$. The scaled version of the graph for the almost sure limit of the fourth spike $\tau_4 = 0.5$

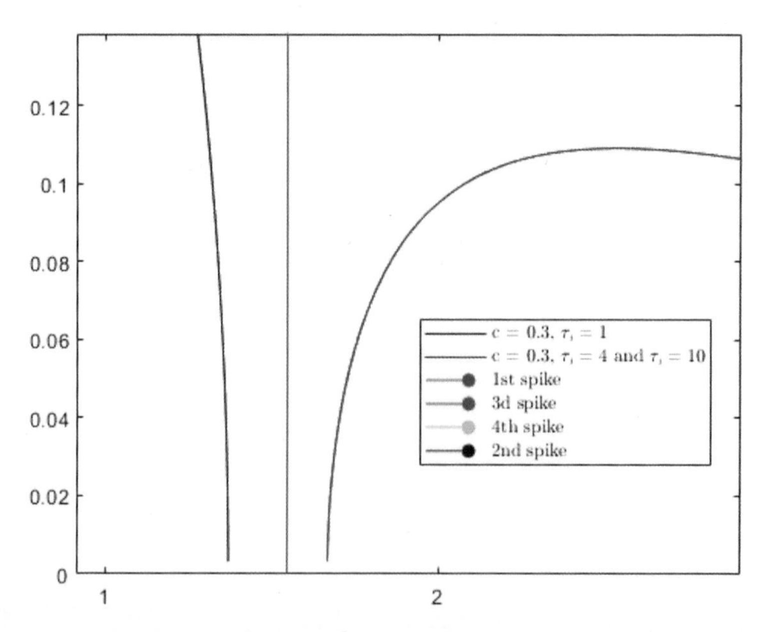

Fig. A.9 The density function for the limiting spectral distribution $F_{c,H}$ of the sample eigenvalues, where $c = 0.3$ and H is uniform on the set $\{1, 4, 10\}$. The scaled version of the graph for the almost sure limit of the third spike $\tau_3 = 2$

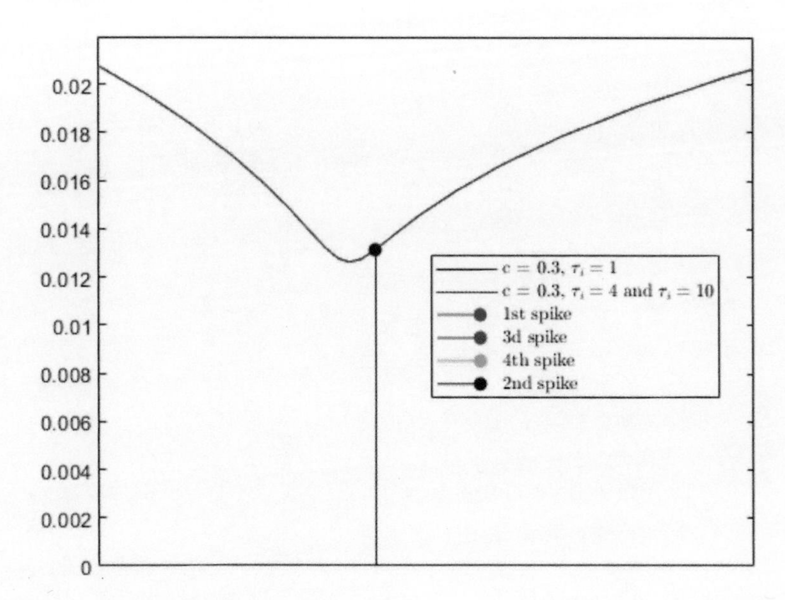

Fig. A.10 The density function for the limiting spectral distribution $F_{c,H}$ of the sample eigenvalues, where $c = 0.3$ and H is uniform on the set $\{1, 4, 10\}$. The scaled version of the graph for the almost sure limit of the second spike $\tau_2 = 6$

Appendix B
Tables

See Tables B.1, B.2, B.3, B.4, B.5, B.6, B.7, B.8, B.9, B.10, and B.11.

Table B.1 Sample covariance matrix. Amplification of the estimation noise. $n = 100$

	c_n	$\left\|\widehat{S}_n - \Sigma_n\right\|_F^2$	$\left\|\widehat{S}_n - \Sigma_n\right\|_1^2$	$\left\|\widehat{S}_n - \Sigma_n\right\|$
p = 30, n = 100	0.3	35.43	63.11	2.53
p = 60, n = 100	0.6	134.78	242.16	4.14
p = 90, n = 100	0.9	299.76	544.44	5.43

Note: The amplification of the estimation noise of the sample covariance matrix in the high-dimensional setup. The true population covariance matrix is $\Sigma_n = \text{diag}(\tau_{n,1}, \ldots, \tau_{n,p})$, $\tau_{n,i} = H_n^{-1}((i - 0.5)/p)$, $H = 1 + 10Z$, and $Z \sim Beta(1, 10)$. The entries of the table are the average values over $R = 10{,}000$ replications

Table B.2 Sample covariance matrix. Amplification of the estimation noise. $n = 1000$

	c_n	$\left\|\widehat{S}_n - \Sigma_n\right\|_F^2$	$\left\|\widehat{S}_n - \Sigma_n\right\|_1^2$	$\left\|\widehat{S}_n - \Sigma_n\right\|$
p = 300, n = 1000	0.3	326.16	564.74	2.87
p = 600, n = 1000	0.6	1297.19	2269.07	4.41
p = 900, n = 1000	0.9	2932.00	5062.65	5.69

Note: The amplification of the estimation noise of the sample covariance matrix in the high-dimensional setup. The true population covariance matrix is $\Sigma_n = \text{diag}(\tau_{n,1}, \ldots, \tau_{n,p})$, $\tau_{n,i} = H_n^{-1}((i - 0.5)/p)$, $H = 1 + 10Z$, and $Z \sim Beta(1, 10)$. The entries of the table are the average values over $R = 10{,}000$ replications

A. Zagidullina, *High-Dimensional Covariance Matrix Estimation*,
SpringerBriefs in Applied Statistics and Econometrics,
https://doi.org/10.1007/978-3-030-80065-9

Table B.3 Sample covariance matrix. Traditional asymptotics

	c_n	$\left\|\widehat{S}_n - \Sigma_n\right\|_F^2$	$\left\|\widehat{S}_n - \Sigma_n\right\|_1^2$	$\left\|\widehat{S}_n - \Sigma_n\right\|$
p = 30, n = 100	0.3	35.36	63.02	2.64
p = 30, n = 500	0.06	7.05	12.35	1.10
p = 30, n = 1000	0.03	3.53	6.19	0.77
p = 30, n = 10000	0.003	0.35	0.62	0.24

Note: The consistency property of the sample covariance matrix under the traditional asymptotics. The true population covariance matrix is $\Sigma_n = \text{diag}(\tau_{n,1}, \dots, \tau_{n,p})$, $\tau_{n,i} = H_n^{-1}((i-0.5)/p)$, $H = 1 + 10Z$, and $Z \sim Beta(1, 10)$. The entries of the table are the average values over $R =$ 10,000 replications

Table B.4 Inverse of sample covariance matrix. Amplification of the estimation error. $n = 100$

	c_n	$\left\|\widehat{S}_n^{-1} - \Sigma_n^{-1}\right\|_F$	$\lambda_{max}/\lambda_{min}$
p = 30, n = 100	0.3	3.61	17.56
p = 60, n = 100	0.6	17.08	77.16
p = 90, n = 100	0.9	215.35	1443.81

Note: The amplification of the estimation noise for the inverse of the sample covariance matrix in the high-dimensional setup. The true population covariance matrix is $\Sigma_n = \text{diag}(\tau_{n,1}, \dots, \tau_{n,p})$, $\tau_{n,i} = H_n^{-1}((i-0.5)/p)$, $H = 1 + 10Z$, and $Z \sim Beta(1, 10)$. The entries of the table are the average values over $R = 10{,}000$ replications

Table B.5 Inverse of sample covariance matrix. Traditional asymptotics

	c_n	$\left\|\widehat{S}_n^{-1} - \Sigma_n^{-1}\right\|_F$	$\lambda_{max}/\lambda_{min}$
p = 30, n = 100	0.3	3.61	17.57
p = 30, n = 500	0.06	0.94	6.72
p = 30, n = 1000	0.03	0.62	5.89
p = 30, n = 10000	0.003	0.19	5.16

Note: The reduction of the estimation noise for the inverse of the sample covariance matrix in the traditional setup. The true population covariance matrix is $\Sigma_n = \text{diag}(\tau_{n,1}, \dots, \tau_{n,p})$, $\tau_{n,i} =$ $H_n^{-1}((i-0.5)/p)$, $H = 1 + 10Z$, and $Z \sim Beta(1, 10)$. The entries of the table are the average values over $R = 10{,}000$ replications

Table B.6 Inconsistency of the sample eigenvalues in the high-dimensional setup, $c_n = 0.2$, $p = 200$, $n = 1000$

τ_1	$\lambda_{n,1}$	τ_2	$\lambda_{n,2}$	τ_p	$\lambda_{n,p}$
2.00	2.43	1.50	2.14	1.00	0.3
	(0.0759)		(0.0335)		(0.0078)
	$\cos\phi(u_{n,1}, v_1)$		$\cos\phi(u_{n,2}, v_2)$		$\cos\phi(u_{n,p}, v_p)$
	0.79		0.31		0.06
	(0.0373)		(0.1705)		(0.0451)

Note: The distribution of random variables comprising the $n \times p$ data matrix Y_n is real Gaussian with mean zero and covariance matrix Σ. The true covariance matrix is computed according to the eigendecomposition $\Sigma = V \cdot T \cdot V'$, where V denotes the $(p \times p)$ orthogonal matrix of eigenvectors, and $T = \text{diag}(\tau_1, \ldots, \tau_p)$ is a diagonal matrix containing the eigenvalues $\tau_i = \frac{i+1}{i}$ (for $i = 1, \ldots, p$). The numbers indicate the mean over the $R = 1000$ Monte Carlo replications, and the numbers in parentheses indicate the corresponding standard errors. v_i is a true i-th eigenvector, and $u_{n,i}$ denotes the estimated i-th sample eigenvector

Table B.7 Inconsistency of the sample eigenvalues in the high-dimensional setup, $c_n = 0.5$, $p = 200$, $n = 400$

τ_1	$\lambda_{n,1}$	τ_2	$\lambda_{n,2}$	τ_p	$\lambda_{n,p}$
2.00	3.07	1.50	2.91	1.00	0.09
	(0.0931)		(0.0505)		(0.0054)
	$\cos\phi(u_{n,1}, v_1)$		$\cos\phi(u_{n,2}, v_2)$		$\cos\phi(u_{n,p}, v_p)$
	0.50		0.14		0.06
	(0.1595)		(0.1003)		(0.0430)

Note: The distribution of random variables comprising the $n \times p$ data matrix Y_n is real Gaussian with mean zero and covariance matrix Σ. The true covariance matrix is computed according to the eigendecomposition $\Sigma = V \cdot T \cdot V'$, where V denotes the $(p \times p)$ orthogonal matrix of eigenvectors, and $T = \text{diag}(\tau_1, \ldots, \tau_p)$ is a diagonal matrix containing the eigenvalues $\tau_i = \frac{i+1}{i}$ (for $i = 1, \ldots, p$). The numbers indicate the mean over the $R = 1000$ Monte Carlo replications, and the numbers in parentheses indicate the corresponding standard errors. v_i is a true i-th eigenvector, and $u_{n,i}$ denotes the estimated i-th sample eigenvector

Table B.8 Properties of sample eigenvalues. Spiked covariance matrix

$c_n = 0.2$ ($p = 200, n = 1000$)				
$\tau_1 = 2$	$\tau_1 = 3$	$\tau_1 = 4$	$\tau_1 = 5$	$\tau_1 = 10$
		Mean($\lambda_{n,1}$)		
2.40	3.31	4.27	5.25	10.25
(0.0791)	(0.1296)	(0.1809)	(0.2270)	(0.4443)
		Theoretical S.E.s		
(0.08)	(0.1308)	(0.1769)	(0.2222)	(0.4467)
		$\widehat{\text{Bias}}(\lambda_{n,1})$		
0.40	0.31	0.27	0.25	0.25
		Theoretical Bias($\lambda_{n,1}$)		
0.40	0.30	0.27	0.25	0.22
		$\tau_2 = 1 + \sqrt{c_n}$		
		Mean($\lambda_{n,2}$)		
2.08	2.08	2.08	2.09	2.09
(0.0314)	(0.0319)	(0.0322)	(0.0321)	(0.0325)
		$\tau_3 = 1$		
		Mean($\lambda_{n,3}$)		
2.03	2.03	2.03	2.03	2.03
(0.0236)	(0.0226)	(0.0226)	(0.0226)	(0.0224)
		$\tau_p = 1$		
		Mean($\lambda_{n,p}$)		
0.31	0.31	0.31	0.31	0.31
(0.0076)	(0.0076)	(0.0078)	(0.0077)	(0.0075)

Note: $p = 200$, $n = 1000$, $c_n = 0.2$. The population covariance matrix is $\Sigma = \text{diag}(\tau_1, \tau_2, 1 \ldots, 1)$
τ_1 takes values in the set $\{2, 3, 4, 5, 10\}$, $\tau_2 = 1 + \sqrt{c_n}$ and $\tau_3 = \ldots = \tau_p = 1$. The number of Monte Carlo replications is $R = 1000$
Remark: for $c_n = 0.2$ the limiting values are $(1 + \sqrt{c_n})^2 = 2.0944$ and $(1 - \sqrt{c_n})^2 = 0.3056$, respectively

Table B.9 Properties of sample eigenvalues. Spiked covariance matrix

$c_n = 0.5$ $(p = 200, n = 400)$				
$\tau_1 = 2$	$\tau_1 = 3$	$\tau_1 = 4$	$\tau_1 = 5$	$\tau_1 = 10$
		Mean($\lambda_{n,1}$)		
3.03	3.75	4.67	5.63	10.52
(0.0938)	(0.1916)	(0.2806)	(0.3449)	(0.6877)
		Theoretical S.E.s		
(0.1000)	(0.1984)	(0.2749)	(0.3480)	(0.7049)
		$\widehat{\text{Bias}}(\lambda_{n,1})$		
1.03	0.75	0.67	0.63	0.52
		Theoretical Bias($\lambda_{n,1}$)		
1.00	0.75	0.67	0.63	0.56
		$\tau_2 = 1 + \sqrt{c_n}$		
		Mean($\lambda_{n,2}$)		
2.86	2.89	2.89	2.89	2.89
(0.0580)	(0.0615)	(0.0635)	(0.0630)	(0.0618)
		$\tau_3 = 1$		
		Mean($\lambda_{n,3}$)		
2.77	2.79	2.79	2.79	2.79
(0.0421)	(0.0444)	(0.0451)	(0.0433)	(0.0456)
		$\tau_p = 1$		
		Mean($\lambda_{n,p}$)		
0.09	0.09	0.09	0.09	0.09
(0.0053)	(0.0052)	(0.0055)	(0.0051)	(0.0053)

Note: $p = 200$, $n = 400$, $c_n = 0.5$. The population covariance matrix is $\Sigma = \text{diag}(\tau_1, \tau_2, 1 \ldots, 1)$.
τ_1 takes values in the set $\{2, 3, 4, 5, 10\}$, $\tau_2 = 1 + \sqrt{c_n}$ and $\tau_3 = \ldots = \tau_p = 1$. The number of Monte Carlo replications is $R = 1000$
Remark: for $c_n = 0.5$ the limiting values are $(1 + \sqrt{c_n})^2 = 2.9142$ and $(1 - \sqrt{c_n})^2 = 0.0858$, respectively

Table B.10 Properties of sample eigenvectors. Spiked covariance matrix

$c_n = 0.2$ ($p = 200, n = 1000$)				
$\tau_1 = 2$	$\tau_1 = 3$	$\tau_1 = 4$	$\tau_1 = 5$	$\tau_1 = 10$
		$\cos\phi(u_{n,1}, e_1)$		
0.81	0.93	0.96	0.97	0.99
(0.0314)	(0.0091)	(0.0050)	(0.0037)	(0.0014)
		$\tau_2 = 1 + \sqrt{c_n}$		
		$\cos\phi(u_{n,2}, e_2)$		
0.31	0.31	0.31	0.31	0.31
(0.1656)	(0.1633)	(0.1656)	(0.1705)	(0.1614)
		$\tau_3 = 1$		
		$\cos\phi(u_{n,3}, e_3)$		
0.05	0.06	0.05	0.05	0.06
(0.0407)	(0.0416)	(0.0401)	(0.0404)	(0.0436)
		$\tau_p = 1$		
		$\cos\phi(u_{n,p}, e_p)$		
0.06	0.06	0.06	0.06	0.06
(0.0427)	(0.0421)	(0.0434)	(0.0434)	(0.0424)

Note: $p = 200$, $n = 1000$, $c_n = 0.2$. The population covariance matrix is $\Sigma = \text{diag}(\tau_1, \tau_2, 1 \dots, 1)$
τ_1 takes values in the set $\{2, 3, 4, 5, 10\}$, $\tau_2 = 1 + \sqrt{c_n}$ and $\tau_3 = \dots = \tau_p = 1$. The number of Monte Carlo replications is $R = 1000$. The value of the cosine converging to 1 means that the angle is converging to 0. $e_1, \dots, e_p$ are the basis vectors

Table B.11 Properties of sample eigenvectors. Spiked covariance matrix

$c_n = 0.5$ ($p = 200, n = 400$)				
$\tau_1 = 2$	$\tau_1 = 3$	$\tau_1 = 4$	$\tau_1 = 5$	$\tau_1 = 10$
		$\cos\phi(u_{n,1}, e_1)$		
0.50	0.83	0.90	0.93	0.97
(0.1731)	(0.0262)	(0.0148)	(0.0097)	(0.0039)
		$\tau_2 = 1 + \sqrt{c_n}$		
		$\cos\phi(u_{n,2}, e_2)$		
0.24	0.29	0.31	0.31	0.31
(0.1504)	(0.1571)	(0.1636)	(0.1622)	(0.1595)
		$\tau_3 = 1$		
		$\cos\phi(u_{n,3}, e_3)$		
0.05	0.06	0.06	0.05	0.05
(0.0397)	(0.0419)	(0.0414)	(0.0405)	(0.0397)
		$\tau_p = 1$		
		$\cos\phi(u_{n,p}, e_p)$		
0.06	0.06	0.06	0.06	0.06
(0.0436)	(0.0435)	(0.0429)	(0.0420)	(0.0426)

Note: $p = 200, n = 400, c_n = 0.5$. The population covariance matrix is $\Sigma = \text{diag}(\tau_1, \tau_2, 1 \ldots, 1)$ τ_1 takes values in the set $\{2, 3, 4, 5, 10\}$, $\tau_2 = 1 + \sqrt{c_n}$ and $\tau_3 = \ldots = \tau_p = 1$. The number of Monte Carlo replications is $R = 1000$. The value of the cosine converging to 1 means that the angle is converging to 0. $e_1, \ldots, e_p$ are the basis vectors

Appendix C
Additional Theoretical Results

C.1 Spiked Population Model

Here some additional results for the Sect. 4.2.4 are presented.

It should be noticed that in $c = 1$ and $c > 1$ cases the spikes are necessarily larger than 1.

Corollary C.1.1 *Let $p \to \infty$, $n = n(p) \to \infty$, $c_n = p/n \to c$, $c = 1$. Let us consider the large fundamental spikes such that $\tau_i > 2$. Then, for the corresponding sample eigenvalues $\lambda_{n,i}$ holds the following:*

(i) Large fundamental spikes $\tau_i > 2$:

$$\lambda_{n,i} \to \tau_i \left(1 + \frac{1}{\tau_i - 1}\right) \; a.s. \text{ for } i \in J_i$$

(ii) Non-fundamental spikes $1 < \tau_i \leq 2$:

$$\lambda_{n,i} \to 4 \; a.s. \text{ for } i \in J_i$$

(iii)

$$\lambda_{n,\min\{n,p\}} \to 0 \; a.s.$$

It should be noticed here once more that the sample eigenvalues $\lambda_{n,i}$ corresponding to the population eigenvalues, $\tau_i = 1$, are distributed according to the Marčenko–Pastur law, where the left and right edges converge almost surely to $(1 - \sqrt{c})^2$ and $(1 + \sqrt{c})^2$, respectively.

A. Zagidullina, *High-Dimensional Covariance Matrix Estimation*,
SpringerBriefs in Applied Statistics and Econometrics,
https://doi.org/10.1007/978-3-030-80065-9

Furthermore, in the special case of $c = 1$, the density function of the Marčenko–Pastur distribution is of the following form:

$$f_1(x) = \frac{1}{2\pi x}\sqrt{x(4-x)}\mathbb{1}\{0 < x \leq 4\}.$$

The bound results in Corollary C.1.1 illustrate that left-edge sample eigenvalue $\lambda_{n,\min\{n,p\}}$ converges almost surely to zero. The right-edge sample eigenvalue corresponding to the cluster of unit population eigenvalues, $\tau_i = 1$, converges almost surely to 4.

Moreover, the sample eigenvalues corresponding to the small non-fundamental spikes, $1 < \tau_i \leq 2$, as well converge to 4. The latter means that the sample eigenvalues associated with the non-fundamental spikes, $1 < \tau_i \leq 2$, are not distinguishable from the right edge of the bulk.

This is well-known in the literature mixture effect when the sample eigenvalues associated with distinct population clusters are blurred together in the high-dimensional setup (see, e.g., El Karoui 2008). This phenomenon arises because there is not enough observations n compared to the dimension p in order to recover the spectral separation of the sample eigenvalues in the limit (see, e.g., Mestre 2008a).

Corollary C.1.2 *Let $p \to \infty$, $n = n(p) \to \infty$, $c_n = p/n \to c$, $c > 1$. Let us consider the large fundamental spikes such that $\tau_i > 1 + \sqrt{c}$. Then, for the corresponding sample eigenvalues λ_i holds the following:*

(i) Large fundamental spikes $\tau_i > 1 + \sqrt{c}$:

$$\lambda_{n,i} \to \tau_i \left(1 + \frac{c}{\tau_i - 1}\right) \ a.s. \text{ for } i \in J_i$$

(ii) Non-fundamental spikes $1 < \tau_i \leq 1 + \sqrt{c}$:

$$\lambda_{n,i} \to (1+\sqrt{c})^2 \ a.s. \text{ for } i \in J_i$$

(iii)

$$\lambda_{n,n} \to (1-\sqrt{c})^2 \ a.s.$$

(iv)

$$\lambda_{n,n+1} = \ldots = \lambda_{n,p} = 0$$

In the case $c > 1$ as well as in case $c \leq 1$, the limiting spectral distribution of the sample eigenvalues corresponding to the unit population eigenvalues, $\tau_i = 1$, is the Marčenko–Pastur law, with the only difference that the left edge of its support truncates at the n-th sample eigenvalue $\lambda_{n,n}$. The almost sure limit of $\lambda_{n,n}$ is equal to $(1-\sqrt{c})^2$. Sample eigenvalues $\lambda_{n,n+1} = \ldots = \lambda_{n,p} = 0$ due to the rank deficiency of the sample covariance matrix $\widehat{S}_n$.

As well as in Corollary C.1.1 the sample eigenvalue corresponding to the right edge of the bulk blurs together with the sample eigenvalues corresponding to the small non-fundamental spikes, $1 < \tau_i \leq 1 + \sqrt{c}$.

C.2 Generalized Spiked Population Model: *Silverstein Equation*

Here some additional results for Sect. 4.2.5 are presented.

The function ψ used in the generalized spiked covariance model is based on the alternative representation of the fundamental Marčenko–Pastur equation (see Proposition 4.2.5).

Let us consider the *companion* sample covariance matrix defined as $\underline{\widehat{S}_n} = \frac{1}{n} Y_n Y_n'$ (Yao et al. 2015), which has the size $n \times n$.

The sample covariance matrix $\widehat{S}_n$ and the *companion* one $\underline{\widehat{S}_n}$ share the same non-zero eigenvalues; thus, their empirical spectral distributions have the following property:

$$nF^{\underline{\widehat{S}_n}} - pF^{\widehat{S}_n} = (n-p)\delta_0,$$

where δ_0 denotes the probability mass at point zero.

Hence, when $c_n = p/n \to c > 0$ as $p, n \to \infty$, for the corresponding limiting spectral distributions $\underline{F}_{c,H}$ and $F_{c,H}$, the following holds true:

$$\underline{F}_{c,H} - cF_{c,H} = (1-c)\delta_0.$$

Moreover, $F^{\widehat{S}_n}$ has limit $F_{c,H}$ if and only if $F^{\underline{\widehat{S}_n}}$ has a limit $\underline{F}_{c,H}$.

Furthermore, the respective Stieltjes transforms $\underline{m}$ and m for $\underline{F}_{c,H}$ and $F_{c,H}$ are linked by the relation:

$$\underline{m}(z) = -\frac{1-c}{z} + cm(z), \ z \in \mathbb{C}^+.$$

Substituting $\underline{m}$ for m into Marčenko–Pastur equation and solving for z yield:

$$g_{c,H}(z) = z = -\frac{1}{\underline{m}} + c \int \frac{t}{1 + t\underline{m}} dH(t).$$

The latter equation is called the *Silverstein equation* for historical reasons though it is equivalent to the Marčenko–Pastur one. This form of the fundamental equation is used in the lemma by Silverstein and Choi (1995).

C.3 Generalized Spiked Population Model: *Multiplicities of Spikes $m_i(\tau_i) > 1, \ i = 1, \ldots, k$*

Here some additional results for Sect. 4.2.5 are presented.

Under generalized spiked population model, let us assume that the population covariance matrix is of the following form:

$$\Sigma_n = \begin{pmatrix} \Sigma_\tau & 0 \\ 0 & \Sigma_{n,\beta} \end{pmatrix},$$

where Σ_τ corresponds to the sub-matrix associated with the fundamental spikes, τ_i's, while the sub-matrix $\Sigma_{n,\beta}$ corresponds to the base population eigenvalues, i.e., $\beta_{n,j}$'s, and the spectrum of the $\Sigma_{n,\beta}$ sub-matrix converges to a non-random limiting spectral distribution H.

Let τ_i be a fundamental spike eigenvalue. The m_i packed sample eigenvalues $\{\lambda_{n,i}, \ i \in J_i\}$ converge almost surely to $\psi_i = \psi(\tau_i)$. Furthermore, the eigendecomposition of Σ_τ is given as following: $\Sigma_\tau = V_\tau \ \mathrm{diag}(\tau_1 \cdot I_{m_1}, \ldots, \tau_k \cdot I_{m_k}) \ V_\tau'$, where V_τ is the orthogonal matrix.

Proposition C.3.1 *Under generalized spiked population model, assume that conditions defined thereafter are valid and the random variables $\{X_{ij}\}$ are real valued. Let τ_i be a fundamental spike eigenvalue of multiplicity m_i and $\{\lambda_{n,i}, \ i \in J_i\}$ be the corresponding spiked sample eigenvalues tending to $\psi_i = \psi(\tau_i)$. Then the m_i-dimensional random vectors*

$$\sqrt{n}(\lambda_{n,i} - \psi_i), \ i \in J_i$$

weakly converge to the distribution of eigenvalues of the $m_i \times m_i$ random matrix

$$M_i = V_{\tau,i}' \ G(\psi_i) \ V_{\tau,i},$$

where $V_{\tau,i}$ is the i-th block of size $p \times m_i$ in V_τ corresponding to the spike τ_i, and $G(\psi_i)$ is a Gaussian random matrix with independent entries such that:

1. *Its diagonal elements are i.i.d. Gaussian, with mean 0 and variance*

$$\sigma^2(\tau_i) = \tau_i^2 \psi'(\tau_i)(2 + \beta_x \psi'(\tau_i)),$$

where $\beta_x = (E|X_{ii}|^4 - 3)$ denotes the fourth cumulant of the base entries X_{ij}.

2. *Its upper triangular elements are i.i.d. Gaussian, with mean 0 and variance*

$$s^2(\tau_i) = \tau_i^2 \psi'(\tau_i).$$

In particular:

1. *When the base entries* $\{X_{ij}\}$ *are Gaussian,* $\beta_x = 0$*, and then* $\sigma^2(\tau_i) = 2s^2(\tau_i)$*, thus, the matrix* G *is a real Gaussian Wigner matrix.*
2. *When the spike* τ_i *is simple, that is,* $m_i = 1$*, the limiting distribution of* $\sqrt{n}(\lambda_{n,i} - \psi_i)$ *is Gaussian.*

It is worth noticing that in case Σ_τ is diagonal, the corresponding orthogonal matrix is identity, $V_\tau = I$, and hence, the joint distribution of the i-th packed spiked sample eigenvalues $\{\lambda_{n,i},\ i \in J_i\}$ is given by the eigenvalues of the Gaussian matrix G. This joint distribution is non-Gaussian unless the spike eigenvalue τ_i is simple, that is, $m_i = 1$.

C.4 Partial Sphericity Test

Here some additional results for Sect. 4.7 are presented.

Proposition C.4.1 *Let* $n\widehat{S}_n = nY_n'Y_n \sim \mathcal{W}_p(\Sigma_n, n)$ *and let us assume the above condition fulfilled. Then, under the null hypothesis* H_k *that the true number of spikes is* k *and fixed,*

$$H_k : \tau_{k+1} = \ldots = \tau_p,$$

the asymptotic distribution of LRT_k *(when* $n, p \to \infty$ *and* $c_n = p/n \to c > 0$*) is given by*

(ii) case $p > n + k$*:*

$$\frac{\ln LRT_k - \mu_{n,p,k}}{\sigma_{n,p,k}} \xrightarrow{d} \mathcal{N}(0, 1),$$

where

$$\mu_{n,p,k} = \tilde{\mu}^*_{n,p,k} + \ln B^*_{n,p,k} + \ln C^*_{n,p,k} + \ln D^*_{n,p,k},$$

$$\sigma^2_{n,p,k} = -2\left\{\frac{n}{p-k} + \ln\left(1 - \frac{n}{p-k}\right)\right\},$$

with

$$\tilde{\mu}^*_{n,p,k} = -n - \left(p - k - n - \frac{1}{2}\right)\ln\left\{1 - \frac{n}{p-k}\right\},$$

$$B^*_{n,p,k} = \left(1 + \frac{\sum_{i=1}^k \lambda_{n,i} - \sum_{i=1}^k m_i\tau_i}{\sum_{i=k+1}^n \lambda_{n,i}}\right)^{n-k}\left(\frac{n-k}{n}\right)^{n-k},$$

$$C^*_{n,p,k} = \left\{ \frac{\sigma^2(p-k)}{n} \right\}^k,$$

$$D^*_{n,p,k} = \prod_{i=1}^{k} \left(1 + \frac{\tau_i}{\sigma^2} \cdot \frac{n}{p-k-n-1}\right)^{m_i} / \prod_{i=1}^{k} \lambda_{n,i}.$$

Remark 1 Here the random variable $\mu_{n,p,k}$ depends on the true values of $\sigma^2, \tau_1, \ldots, \tau_k$. In order to use the asymptotic distribution of Proposition C.4.1 to test the number of the spikes, we need to replace the true values by consistent estimators.

The parameter σ^2 can be replaced by its consistent estimator

$$\widehat{\sigma}^2 = \sum_{i=k+1}^{p} \lambda_{n,i}/(p-k).$$

Sample eigenvalues $\lambda_{n,i}$ according to Proposition 4.2.6 experience a phase transition in the limit: if $\tau_i > \sigma^2(1+\sqrt{c})$, then

$$\lambda_{n,i} \to \tau_i \left(1 + \frac{c\sigma^2}{\tau_i - \sigma^2}\right),$$

whereas for the population eigenvalues τ_i that are in the interval $(\sigma^2, \sigma^2\left(1+\sqrt{c}\right)]$, the limit is $\sigma^2\left(1+\sqrt{c}\right)^2$.

Thus, the sample eigenvalues $\lambda_{n,i}$ corresponding to the spike population eigenvalues are biased estimators. Given the form of the bias, we can construct the following consistent estimator $\tilde{\tau}_i$ for the population eigenvalues that cross the threshold $\tau_i > \sigma^2(1+\sqrt{c})$:

$$\tilde{\tau}_i = \frac{1}{2}\left\{\lambda_{n,i} + \hat{\sigma}^2 - \hat{\sigma}^2\frac{p}{n} + \sqrt{-4\lambda_{n,i}\hat{\sigma}^2 + \left(\lambda_{n,i} + \hat{\sigma}^2 - \hat{\sigma}^2\frac{p}{n}\right)^2}\right\}.$$

In the limit, the discriminant will be non-negative if and only if $\tau_i > \sigma^2(1+\sqrt{c})$; however, the sample version of the discriminant can be negative when the population eigenvalue τ_i is close to the threshold (or less than the threshold).

In that case the estimator of τ_i is considered to be $\tilde{\tau}_i = \hat{\sigma}^2(1+\sqrt{p/n})$ since that is the value that makes the discriminant equal to zero.

C.5 Passemier and Yao Test Calibration

Here some additional results for Sect. 4.7.2 are presented.

The idea of the calibration procedure for the "tuning" parameter C is to use the difference of the two largest eigenvalues of a Wishart matrix (this corresponds to the case without any spike).

The estimator $\hat{q}_n$ is determined once two consecutive sample eigenvalues $\lambda_{n,j}$ and $\lambda_{n,j+1}$ are below the threshold d_n corresponding to a noise part of sample eigenvalues.

As the distribution of the difference between eigenvalues of a Wishart matrix is not known explicitly, R independent replications can be drawn to approximate the distribution of the difference between the two largest eigenvalues $\lambda_{n,1} - \lambda_{n,2}$ numerically.

The quantile $s(\alpha)$ such that $P(\lambda_{n,1} - \lambda_{n,2} \leq s(\alpha)) = 1 - \alpha$ is then estimated numerically.

Thus, the "tuned" parameter C is set to be equal to:

$$\tilde{C} = s(\alpha) \cdot n^{2/3} / \sqrt{2 \log \log(n)}.$$

References

El Karoui, N. (2008). Spectrum estimation for large dimensional covariance matrices using random matrix theory. *Annals of Statistics, 36*(6), 2757–2790.

Mestre, X. (2008). Improved estimation of eigenvalues and eigenvectors of covariance matrices using their sample estimates. *IEEE Transactions on Information Theory, 54*(11), 5113–5129.

Silverstein, J., & Choi, S. (1995). Analysis of the limiting spectral distribution of large dimensional random matrices. *Journal of Multivariate Analysis, 54*(2), 295–309.

Yao, J., Zheng, S., & Bai, Z. (2015). *Large sample covariance matrices and high-dimensional data analysis*. Cambridge University Press.

Printed in Great Britain
by Amazon